Rabbit Production and Management

AF634422

NIPA® GENX ELECTRONIC RESOURCES & SOLUTIONS P. LTD.
New Delhi-110 034

www.nipaers.com
Browse, Search, Read & Buy...
Print Books
eChapters
eBooks
Publishers
Forthcoming
Subject Catalogues
New Titles
Online Resources on:
✓ Current Affairs
✓ Reasoning, Logic & Aptitude
✓ Competitive Examinations
✓ English Language Lab
✓ Writing & Pronunciation Tools
✓ Personality Development
✓ Online Programmes for Professionals
✓ Interview Preparation
Pay Using
PhonePe
Google Pay
paytm
PayPal
UPI
CC Avenue

Rabbit Production and Management

M. T. Banday
H. P. Shrivastava
Henna Hamadani

NIPA® GENX ELECTRONIC RESOURCES & SOLUTIONS P. LTD.
New Delhi-110 034

NIPA® GENX ELECTRONIC RESOURCES & SOLUTIONS P. LTD.

101,103, Vikas Surya Plaza, CU Block
L.S.C. Market, Pitam Pura, New Delhi-110 034
Ph : +91 11 27341616, 27341717, 27341718
E-mail: newindiapublishingagency@gmail.com
web: www.nipabooks.com

For customer assistance, please contact
Phone: + 91-11-27 34 17 17
Fax: + 91-11- 27 34 16 16
E-Mail: feedbacks@nipabooks.com

ISBN: 978-81-19254-64-4

Composed and Designed by NIPA®.

Sher-e Kashmir University of Agricultural Sciences and Technology of Kashmir

www.skuastkashmir.ac.in

DR. TEJ PARTAP
Vice-Chancellor

Foreword

Rabbit domestication dates back no further than previous millennium. It were the Phoenicians who are said to have discovered the wild rabbit Oryctolagus cuniculus of Southern Europe and North Africa when they reached shores of Spain about 1000 B.C.

Several breeds of rabbits were known in the sixteenth century and the fact is correlated by records available that controlled breeding had been introduced by this time. As it was most delectable dish of the monks, domestication of rabbits became a reality in the Middle Age.

However, it was only in the nineteenth century that rabbit rearing was undertaken in almost all parts of the Western Europe. Although backyard rabbit farming has been practiced in Indian villages for a few decades, the scientific rabbit rearing on commercial lines is relatively a new concept in India.

In India, proven breeds of exotic origin were imported in fifties by private breeders to help the farmers to increase their income through it has been confined to cooler hilly areas of the country such as Jammu & Kashmir, Himachal Pradesh and parts of Uttar Pradesh and of late in North-eastern states as well.

Shalimar, Srinagar-191121, J&K, India. PO Box No. 262, GPO Srinagar-190 001, J&K
Phone: (O) 0194-2462159, 2464028 Fax: 0194-2462160 (R) 0194-2461543, 2463655 Fax: 0194-2461543
E-mail: vc@skuastkashmir.ac.in, skuastkvc@gmail.com

Nevertheless rabbit farming can open vistas of self employment for the young educated youth and help the government mitigate the plight of the marginal farmers as well.

Since rabbit farming is quite a new field in India, no authentic book is available that takes stock of Indian conditions.

The authors have made a sincere attempt to acquaint the farming community and researchers related to this particular field by the way of providing precise information related to production and management of rabbit. The authors have written on the practical aspects of rabbit farming in India. I am sure the book will, indeed, prove a handbook of guidance for the students, researchers and farmers.

(Tej Partap)

Shalimar, Srinagar-191121, J&K, India. PO Box No. 262, GPO Srinagar-190 001, J&K
Phone: (O) 0194-2462159, 2464028 Fax: 0194-2462160 (R) 0194-2461543, 2463655 Fax: 0194-2461543
E-mail: vc@skuastkashmir.ac.in, skuastkvc@gmail.com

Preface

The book provides the general background about production aspects of rabbit rearing. The information provided in the book has been based on the rearing practices adopted for rabbits in India. Keeping in view the relevance of rabbit production, specific topics have been chosen about the rabbits.

The information provided in the book is based on the vast professional experiences, research findings and latest data from unmatched sources. The book has been written in a manner so that it could be used extensively by the students, teachers and academics in the educational and research institutes dealing with Veterinary Sciences, Animal Sciences and Livestock Production. The field functionaries in the Animal Husbandry Department and livestock farms both in public as well as private sector will find the book very useful tool for successful rabbit production.

Experienced rabbit owners can use it as a guide and a reference book and novices in this field can gain tremendous knowledge from this book. The following chapters have been included in the book.

1. Introduction
2. Breeds of rabbit
3. Housing of rabbit
4. Nutrition and feeding
5. Reproduction in rabbit
6. Management of rabbits
7. Diseases of rabbit
8. Angora rabbit production
9. Clinical examination and laboratory procedures
10. Rabbit behavior
11. Slaughter and skinning of rabbits

Simple and lucid illustrations make the book quite a reading pleasure.

Contents

Foreword .... *v*

Preface .... *vii*

1. Introduction **1**

1.1. History and Origins of Rabbits 1

1.2. Special Features about Rabbits 6

1.3. Mechanism of Adaptations in Rabbits 7

2. Breeds of Rabbits **11**

2.1. Classification of Rabbits 11

3. Housing of Rabbits **21**

3.1. Ideal Environmental Conditions Needed for Rearing of Rabbits 21

3.2. Housing Systems 24

3.3. Equipments Required in Cages 33

4. Nutrition of Rabbit **39**

4.1. Digestive System of Rabbit 39

4.2. Mechanism of Digestion 44

4.3. Digestive Tract and Caecatrophy 45

4.4. Feeding Behaviour 47

4.5. Principle Components of the Rabbit Diet 48

4.6. Feed Regimes during Different Life Stages 52

4.7. Feeding Schedule 54

4.8. Feed Consumption 55

4.9. Nutrient Requirement for Rabbits 55

4.10. Feeding of Rabbit 55

4.11. Homemade Feed for Non Breeding Rabbit 56
4.12. Homemade Diet for Pregnant Rabbit 56
4.13. Homemade Diet for Lactating does 56
4.14. Time of Feeding 57
4.15. Composition of Standard Pellet Mixture 57
4.16. Some Consideration on Feeding 57
4.17. Utensil for Feeding 58

5. Reproduction in Rabbits 59

5.1. Anatomy of the Genitals 59
5.2. Male Reproduction Physiology 60
5.3. Female Reproduction Physiology 62
5.4. Fertilization and Gestation 63
5.5. Pseudo Pregnancy 64
5.6. Kindling 64
5.7. Artificial Insemination in Rabbits 64
5.8. Lactation 65
5.9. Aspects of Milk Production 65
5.10. Reproduction and Environment 66
5.11. Rates of Reproduction 67

6. Management of Rabbitry 71

6.1. The Production Cycle 71
6.2. Mating 71
6.3. Determining Pregnancy 72
6.4. Kindling 73
6.5. Fostering 73
6.6. Weaning 74
6.7. Stock Renewal 74

6.8. Castration of Young Male Rabbit 75

6.9. Handling Rabbits 76

6.10. Identification 77

6.11. Record Keeping 79

7. Diseases of Rabbit 81

7.1. Skin Diseases 82

7.2. Diseases of Reproductive System 94

7.3. Diseases of Urinary System 100

7.4. Diseases of Respiratory System 107

7.5. The Digestive System 111

7.6. The Musculoskeletal System 130

8. Angora Rabbit Production 137

8.1. Reproduction 137

8.2. Mating 137

8.3. Gestation 138

8.4. Kindling 138

8.5. Weaning 138

8.6. Sexing 138

8.7. Identification 138

8.8. Harvesting of Wool 139

8.9. Angora Wool Quality 140

8.10. Wool Grading and Marketing 141

9. Clinical Examination and Laboratory Procedures 143

9.1. Handling 143

9.2. Examination 144

9.3. Oral Medication 146

9.4. Injection Techniques 147

9.5. Catheterisation 147
9.6. Pre-Anaesthetic Preparation 147
9.7. Premedication 148
9.8. Sedation 148
9.9. Inhalation Anaesthesia 149
9.10. Induction 150
9.11. Maintenance of Anaesthesia 151
9.12. Endotracheal Intubation 151
9.13. Anaesthetic Monitoring 152
9.14. Post-Operative Care 152
9.15. Surgical Procedures 153

10 Rabbit Behaviour 155

10.1. Territorial Behavior 155
10.2. Social Behavior 156
10.3. Sexual Behavior 156
10.4. Maternal Behavior 156
10.5. Feeding Behavior 157
10.6. Aggressive Behavior 157

11 Slaughter and Skinning of Rabbits 159

11.1. Slaughtering and Dressing 159
11.2. Collection, Preservation and Storage of Pelts 160
11.3. Curing and Glossing of Pelts 162
11.4. Sorting and Grading Pelts 163

Annexures 165

Index 181

Chapter – 1

Introduction

1.1. History and Origins of Rabbits

The existence of rabbit with human being remained from time immemorial. Romans were the first to domesticate wild rabbits in the 5th century by the monks of the Champagne region in France as a readily available food source. Rabbits were introduced to Britain during the 12th Century, and during the Middle Ages, the breeding and farming of rabbits for meat and fur became widespread throughout Europe. Sources suggest that some women among the Medieval gentry even kept rabbits as pets. The selective breeding of European rabbits meant that distinct breeds arose in different regions, and the origins of many old breeds can be traced back several centuries. For example, paintings from the 15th century show rabbits in a variety of colours, some even with white 'Dutch' markings; 16th century writings suggest that the Flemish Giant was already being pure-bred under the name Ghent Giant, in the Flemish speaking Ghent area of Belgium; 17th century sources tell of the arrival of 'silver' rabbits in England and France, brought from India and China by seafarers and influential in the Silver and Champagne de Argente breeds; 18th century sources suggest a breed known as Lapin de Nicard once existed in France and weighed as little as 1.5kg (3½ lbs), some consider this to be the forerunner of all dwarf breeds; the English Lop can also be traced back to 18th century records, and is considered the ancestor of all the lop breeds. By the middle of the 19th century, the widespread practice of selectively breeding domestic rabbits had resulted in a large variety of breeds, ranging from the tiny Polish rabbit to the huge Flemish Giant.

The European wild rabbit have evolved around 4,000 years ago on the Iberian Peninsula, the name 'Hispania' (Spain) is translated from the name given to that area by Phoenician merchants, meaning 'land of the rabbits'. When the Romans arrived in Spain around 200BC, they began to farm the native rabbits for their meat and fur. The Romans called this practice 'Cuniculture' and kept the rabbits in fenced enclosures. Inevitably, the rabbits tried to escape and it is perhaps no surprise that the latin name *'Oryctolagus cuniculus'* means 'hare-like digger of underground tunnels'. The spread of the Roman empire, along with increasing trade between countries, helped to introduce the European rabbit into many more parts of Europe and Asia. With their rapid reproduction rate, and the increasing cultivation of land providing ideal habitat, rabbits soon established large populations in the wild. The European rabbit continued to be introduced to new countries as they were explored, or colonised by European adventurers and pioneers. Wild rabbits thrived in many new locations, and populations grew rapidly in countries with suitable habitat and few natural predators. The European rabbit became widespread in North America and Australia, for example, where the wild rabbit has become a troublesome pest to farmers and conservationists.

Up until the 19th century, domestic rabbits had been bred purely for their meat and fur, but during the Victorian era, many new 'fancy' breeds were developed for the hobby of breeding rabbits for showing. Industrialisation also meant that many people moving from the country to the expanding towns and cities, brought rabbits with them; apart from poultry, they were the only 'farm' animal to be practical to keep in town. Although many of these rabbits were bred for meat, it became increasingly common among the rising middle classes to keep rabbits as pets. Rabbits were connected with the countryside and the animals they had left behind, and became considered almost sentimentally. Rabbit wares were promoted in connection with children, and the romantic attitude towards rabbits persists today in the association of 'bunnies' with newborn babies, and the idea of rabbits as a children's pet. By the 20th century, rabbit breeding had become a popular hobby across Europe, with many rabbit fanciers developing new varieties and colours. Some breeds, such as the Himalayan and Rex, came about as the result of naturally-occuring genetic mutations which were then fixed or enhanced through a selective breeding programme. Others were developed through cross-breeding, particularly with rabbits imported from other countries as a result of increasing travel in Europe. Many breed societies and clubs were established, with some breeds undergoing dramatic swings in popularity, often due to changing fashions for fur and commercial uses. Although the European rabbit arrived in America with european settlers, and established a large wild population, rabbits were mostly hunted in the wild until the late 19th century. Domestic rabbitry did not become popular in the United States until around the turn of the century, when many European breeds began to be imported, and breeders also developed some American breeds.

During the two World Wars, governments in both Britain and the United States encouraged people to keep rabbits as a source of homegrown meat and fur, both for themselves and to help feed and clothe soldiers. After the wars, many people continued to keep rabbits in their gardens, and they become commonplace as household pets. Rabbits have become the third most popular pet after cats and dogs in the UK, unlike cats and dogs however they are traditionally seen as 'Children's pets', and often sadly misunderstood. During the last 30 years or so, attitudes towards rabbits as pets have been undergoing a gradual shift. The promotion of rabbit welfare is fostering a greater understanding of rabbits; from their basic needs to their intelligence, personality and behaviour. Rabbits are increasingly seen in the same way as cats and dogs, as a rewarding companion or family pet, and provided with the same level of care and attention, from routine vaccinations and healthcare, to greater freedom and interaction with their owners.

Rabbits belong to the order of mammals called *Lagomorpha*, which includes 40 or so species of rabbits, hares and Pikas. Fossil records suggest that *Lagomorpha* evolved in Asia at least 40 million years ago, during the Eocene period. The break-up of continents during this period may be responsible for the wide distribution of differing species of rabbits and hares around the world, with the exception of Australia. There are currently more than 60 recognised breeds of domestic rabbit in Europe and America, all of them descended from the European rabbit (*Oryctolagus cuniculus*), the only species of rabbit to have been widely domesticated. It is a seperate species from other native rabbits such as the North American jackrabbits and cottontail rabbits and all species of hares (Table 1.1).

In India, from early sixtees the rabbit farming gained importance in places like Himachal Pradesh and Jammu & Kashmir through the introduction of German Angora breed. In 1977, government of India took much interest in rabbit raising in different states and established a Central Research Station to encourage rabbit farming and to envisage research program to mitigate the problem in rabbit farming.

Table 1.1 : Scientific Classification of Rabbit

Family	*Animalia*
Superphylum	*Chordata*
Phylum	*Vertebrata*
Super class	*Gnathostoma*
Class	*Mammalia*
Sub class	*Theria*
Order	*Lagomorpha*
Family	*Leporidae*
Sub families	**Sub family: *Palaeolaginae***
	Genera:
	Pentalagus (East Asia)
	Pronolagus (Southeast Africa)
	Romerolagus (Mexico; One sole species – *R. nelsoni*)
	Sub family: *Leporinae*
	Genera:
	Lepus (Hares; Throughout Old & New World)
	Macrotolagus (Throughout North & Central America)
	Oryctolagus (True rabbit with few species like *O.cuniculus)*
	*Sylvilagus (*American rabbits)
	Coprolagus (Asiatic rabbits)
	Nesolagus (Sumatra; One sole species: *N. netscheri)*
	Brachylagus (Pygmy rabbit; North America)

Species:

Common name	Scientific name	Alternative names
European Rabbit	*Oryctolagus cuniculus*	Domestic Rabbit, Pet Rabbit
African Savanna Hare	*Lepus victoriae*	
Alaskan Hare	*Lepus othus*	Tundra Hare, Beringian Hare
Amami Rabbit	*Pentalagus furnessi*	
Antelope Jackrabbit	*Lepus alleni*	
Arctic Hare	*Lepus arcticus*	
Black Jackrabbit	*Lepus insularis*	
Black-tailed Jackrabbit	*Lepus californicus*	
Broom Hare	*Lepus castroviejoi*	
Brush Rabbit	*Sylvilagus bachmani*	
Bunyoro Rabbit	*Poelagus marjorita*	
Burmese Hare	*Lepus pequensis*	
Brown Hare	*Lepus capensis*	

Contd.

Chinese Hare	*Lepus sinensis*	
Corsican Hare	*Lepus corsicanus*	
Desert Cottontail	*Sylvilagus audubonii*	Audubon's Cottontail
Dice's Cottontail	*Sylvilagus dicei*	
Eastern Cottontail	*Sylvilagus floridanus*	
Ethiopean Hare	*Lepus fagani*	Abyssinian Hare
Ethiopean Highland Hare	*Lepus starcki*	
European Hare	*Lepus europaeus*	
Granada Hare	*Lepus granatensis*	
Hainan Hare	*Lepus hainanus*	
Hispid Hare	*Caprolagus hispidus*	
Indian Hare	*Lepus nigricollis*	
Jameson's Red Rock Hare	*Pronolagus randensis*	
Japanese Hare	*Lepus bracyurus*	
Korean Hare	*Lepus coreanus*	
Marsh Rabbit	*Sylvilagus palustris*	
Mexican Cottontail	*Sylvilagus cunicularius*	
Mountain Cottontail	*Sylvilagus nuttallii*	
Mountain Hare	*Lepus timidus*	
Natal Red Rock Hare	*Pronolagus crassicaudatus*	
New England Cottontail	*Sylvilagus transitionalis*	Wood Rabbit
Omilteme Cottontail	*Sylvilagus insonus*	
Pygmy Rabbit	*Brachylagus idahoensis*	
Riverine Rabbit	*Bunolagus monticularis*	
San Jose Brush Rabbit	*Sylvilagus mansuetus*	
Scrub Hare	*Lepus saxatilis*	
Smith's Red Rock Hare	*Pronolagus rupestris*	
Sumatra Short Eared Rabbit	*Nesolagus netscheri*	
Snowshoe Hare	*Lepus americanus*	
Swamp Rabbit	*Sylvilagus aquaticus*	Water Rabbit
Tapeti	*Sylvilagus brasiliensis*	
Tehuantepec Jackrabbit	*Lepus flavigularis*	
Tolai Hare	*Lepus tolai*	
Tres Marias Cottontail	*Sylvilagus graysoni*	
Volcano Rabbit	*Romerolagus diazi*	
White-sided Jackrabbit	*Lepus callotis*	
White-tailed Jackrabbit	*Lepus townsendii*	
Woolly Hare	*Lepus oiostolus*	
Yarkand Hare	*Lepus yarkandensis*	
Yunnan Hare	*Lepus comus*	

1.2. Special Features about Rabbits

The special features in rabbits are as following:

- Rabbits are midway between ruminants and monogastric animals. They are called pseudo ruminant. They can convert cellulose into meat and can easily survive on kitchen scrap. They are coprophagous in nature they reinvest their faeces in the early hours of morning directly from their anus through lips. Thus replenish vitamins and cellulose content of feaces.
- They breed eight times in a year.
- Female may produce five to eight youngs which attain sexual maturity in four to six months.
- Life span of rabbit is about seven to eight years.
- Slaughter weight of rabbit is about 2 kg. which can be achieved in 12 to 15 weeks.
- Rabbits have got fast reproductive process and are able to remate within 24 hours of giving birth (kindling).
- Their gestation period ranges to 31 days. They are prolific breeder.
- Rabbits have an efficient feed conversion ratio, which is the amount of feed consumed per pound of gain.
- Rabbit meat is highly nutritious, tasty and excellent in quality. It is highest in protein, lowest in fat, cholesterol and sodium (Table 1.2). Hence recommended for cardiac patients.
- It has the least number of calories per pound and has only 8 percent bone.
- Rabbit meat is especially good for babies, senior citizens and anyone with stomach disorder because it is easily digested.
- Rabbit meat can be prepared in over 300 different ways.
- Unlike wild rabbit, domestic rabbit meat is pearly white, tender, juicy and mild in flavor.

Table 1.2 : Comparison in the Composition of Meat from Different Species

Species	Protein	Fat	Calories/pound
Rabbit	25.5%	10.2%	795
Chicken	21.5%	11.0%	810
Veal (med. fat)	18.8%	14.0%	910
Turkey	20.1%	20.2%	1,190
Lamb (med. fat)	15.7%	27.7%	1,420
Beef	19.0%	28.0%	1,440
Pork (med. fat)	13.3%	45.0%	2,050

1.3. Mechanism of Adaptations in Rabbits

A. **EARS:** The ears of rabbit more than 10 cm long and these help the rabbit listen to the faintest of the sounds made by nearby predators. The peculiar outer shape of their ear, called the pinna, is helpful in detecting sound direction very specifically and accurately. The ears move around like radar dishes always on guard for signs of danger. Rabbit ears contain an abundance of blood vessels which make them the most sensitive area of the body. The ears also function as the air conditioning system as rabbits cannot sweat.

B. **EYES:** Rabbit eyes are big, dark, clear and bright, each having 190 degree field of vision so they can see behind also, but it produces a blind spot in front of the nose. Their vision for distance is acute, and they can see eight times better in the dark than humans. Rabbits spend a good deal of time scanning overhead as well. It is an adaptation to look for predators as most are bigger in size, and many attack from above. Scientists use rabbit eyes for study because of their similarity to human eyes.

C. **TEETH:** The teeth are one of the things that set rabbit apart from a rodent because it has two sets of upper incisors instead of the one set rodents. The front incisors are longer and both sets keep growing all through the rabbit's life. As they are herbivores, these incisors help in snapping off part of the plant. The teeth also allow the rabbits to bite if in danger.

D. **NOSE:** Nose of the rabbit twitches. The faster it twitches the more curious (or frightened) is the rabbit. It picks up the slightest scent because scents are the main means of communication between rabbits. Also, the nose is the main way a bunny breathes.

E. **LIPS:** Perspiration glands are located on the lips of rabbit. The rabbit's harelip helps in the gathering of food. These lips are what tastes and grabs the food.

F. **JAWS:** The lower jaw is smaller than the upper one. The jaws move sideways in order for the molars to line up and they average 120 chews a minute.

G. **WHISKERS:** Whiskers are very important in acting as feelers in finding their way, and in helping locate food.

H. **VOCAL CHORDS:** Rabbits can grunt, squeak, and hiss. They can also scream when they are very frightened (they can die from fright) or are experiencing intense pain.

I. **SKULL**: Is unique in that it has a joint located in the back of the head that allows the skull to absorb the shock that comes from the rabbit's jumping and twisting at high speeds.

J. **DEWLAP:** The loose skin area under the neck of the doe. Does will pull fur from here when making a nest to have a litter. Does are usually larger than bucks.

K. **LEGS:** The hind legs of rabbits are long and very strong. These help them hop at a very fast rate. The feet are padded with fur to dampen the shock of landing. European rabbits are known to run especially fast at 16 m/s. They can also change directions relatively quickly to confuse the chasing predator. Rabbits also have sharp toe nails that help in gripping the ground as well as making burrows. The hind legs are so strong that they are also used to kick when in danger.

L. **DIGESTION:** Most of the rabbit's digestion takes place in the large intestine and the caecum. The latter is ten times the size of the stomach. Because of their largely cellulose diet, the caecum has a muscular structure that helps them differentiate between the digestive and the fibrous material. Rabbit are also known to defecate the contents of the caecum, known more commonly as soft pellets, and then eat them again to reabsorb the nutrients. This helps them gather as much nutrients as possible from the relatively hard to digest plant parts.

M. **FUR COLOUR:** Mostly pastel shades of brown, grey and buff, depending on where the rabbits live. While arctic rabbits have a whitish fur colour to camouflage the white background of winter, but in autumn the same colour changes to brown or reddish to mingle with the surroundings. European rabbits have a colour similar to the shades of the ground too.

N. **REPRODUCTION:** Rabbits reproduce at a very fast rate, with the mating period being as long as nine months of a year. A single female can deliver up to 4 to 8 babies at one time, and because of the speed with which the babies are weaned, the female is ready to deliver another sets of kits or baby rabbits in no time. In fact in a year a single family can grow to staggering 800 rabbit members if all conditions are suitable. The rapid reproduction also helps the rabbits adapt to environmental changes faster than many other species.

Chapter – 2

Breeds of Rabbit

Domestication of the rabbit is relatively recent and most of today's breeds have been bred by man in the last 200 to 300 years.

The rabbit has been used as an experimental animal in genetics, reproduction and physiology since the beginning of the century, but it was not until 1950 that the first findings on quantitative genetics were published in Venge's study of maternal influence on rabbit birth weight.

Rabbit breed or populations can also be defined in terms of gene frequencies. Research on rabbit breeding behavior and production development started only recently less than 20 years ago though formal genetic research has a longer history. For production and selection this can be both an advantage and a drawback. The advantage is that there is less temptation for countries examining their own problems. A relative drawback is that the newness of this field of research makes it necessary to establish the appropriate genetic improvement pattern for each region or country.

Various breeds are available for developing or modernizing production. In most developing countries, highly intensive production with complete control over technical and environmental detail is precisely the opposite of what is needed.

2.1. Classification of Rabbits

Rabbit breeds are generally classified in five different types depending upon the:

1. Type of fur/hair
2. Type of body

3. Size of body
4. Purpose of rearing
5. Rabbit tameness

Each classification group will sometimes be referred to as a "**Type of Rabbits**".

2.1.1. Classification of rabbits according to the type of body

One of the most popular ways to classify rabbit breeds is by the type of body. There are five different rabbit body types:

A. Commercial
B. Compact
C. Cylindrical
D. Full-Arch
E. Semi-Arch

A. Commercial body type

Rabbits which are classified under **commercial rabbit body type** have a gradual rise and fall up to the hips upto its tail. Rabbits in this category are generally very good meat producing rabbits due to their large loins and rapid growth rates. Some breeders have pointed out that these commercial breeds seem to look like a larger version of the "Compact Rabbits".

The following rabbit breeds have been recognized under this category:

List of Breeds			
American Chinchilla	**American Sable**	**Californian**	**Champagne d' Argent**
Cinnamon	Creme d' Argent	French Angora	French Lop
Giant Angora	Harlequin	Hotot	New Zealand
Palomino	Rex	Satin	Satin Angora
Silver Fox	Silver Marten		

B. Compact body type

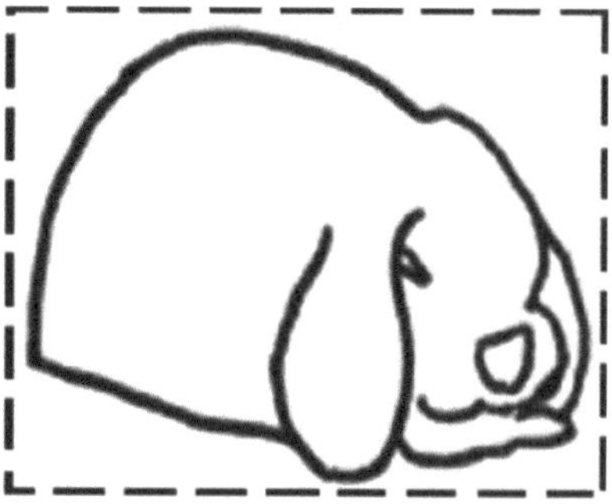

The small and medium rabbit breeds have a **compact body.** These rabbits generally will look balanced when placed properly. When you run your hand over their bodies, your hand should gradually rise up across the rabbit to the hips and then curve itself smoothly down the rabbit to its tail.

The following rabbit breeds have been recognized under this category:

List of Breeds

American Fuzzy Lop	Chinchilla	Dutch	Dwarf Hotot
English Angora	Florida White	Havana	Holland Lop
Jersey Wooly	Lilac	Mini Lop	Mini Rex
Mini Satin	Netherland Dwarf	Polish	Silver
Standard Chinchilla	Thrianta		

C. Cylindrical body type

There is only one classified **cylindrical rabbit breed**. This group is made up of solely the Himalayan Rabbit. The Himalayan rabbits has the appearance of a Californian however, it is much smaller and of course cylindrical. When you judge this rabbit breed you should lay the animal out on the table in a "cylindrical" like fashion. See the picture above for a visual image. Eg. Himalayan rabbits

D. Fully arched body type

The **fully arched rabbit body type** is known for categorizing rabbits which sit in an "always alert" like fashion and which seem to have a very energetic personality. These rabbits have straight erect ears and several have a spotted coat.

The following breeds have been recognized under this category:

List of Breeds			
Belgian Hare	Britannia Petite	Checkered Giant	English Spot
Rhinelander	Tan		

E. Semi-arched body type

Rabbits in this group are large with low shoulders and a high hip. Some rabbit raisers will refer to this group as the "**Gentle Giants**".

The following breeds have been recognized under this category:

List of Breeds				
American	Beveren	English Lop	Giant Chinchilla	Flemish Giant

2.1.2. Classification according to types of rabbit fur

Rabbits are classified based on the type of fur as under:

A. Normal fur

B. Rex fur

C. Satin fur

D. Angora- Wool fur

E. Multi colour fur

A. Normal fur

Most rabbits are classified as having "normal fur". There are actually two different types of normal fur;

a) Flyback

b) Rollback

a) Flyback fur

When a rabbit has flyback fur its coat will "fly back" into place after being petted backwards. The following breeds have been recognized under flyback fur:

List of Breeds

New Zealand	Californian	Dutch	Polish
Palomino	Silver	American	Cinnamon
Havana	English Lop	French Lop	Himalayan
Harlequin	Tan	Belgian Hare	English Spot
Florida White	Rhinelander	Silver Marten	Giant Chinchilla
Britannia Petite	Checkered Giant	Champagne	Creme D' Argent

b) Rollback fur

Rollback fur is longer than flyback fur and will "roll back" into place after running your hand against it. The following breeds have been recognized under rollback fur:

List of Breeds

Satin	Lilac	Hotot	Mini Lop	Holland Lop
Flemish Giant	Beveren	American Sable	Dwarf Hotot	Netherland Dwarf
Silver Fox	American	Standard Chinchilla		

B. Rex fur

Rex fur is of a soft and velvety texture. The fur has a velvety touch and is short. It is used also to make fine fur coats and hats. Rex Fur is a fairly rare type of rabbit fur and is only found in two rabbit breeds; the Rex and the Mini Rex.

C. Satin fur

Satin fur is much softer than normal fur. The fur has a special shine to it. It is used to make fine fur coats and hats. There are only two breeds which possess "Satin Fur". These breeds include the Satin and Mini Satin.

D. Wool fur / Angora fur

Wool fur is much like that of sheep fur however it is proven to be much more dense. The fur grows very long and is shaved or plucked and woven into yarn to be used in making Angora sweaters, hats, and mittens. The long haired rabbits can become a real problem for the casual rabbit breeder. Their fur has a tendency to matt and shed. It's a real bear trying to clean the cages that have long hair everywhere. The hair also floats out and settles on anything in your rabbitry. There are six rabbit breeds which possess "Wool Fur". These breeds include the; American Fuzzy Lop, English Angora, French Angora, Giant Angora, Jersey Wooly and Satin Angora rabbits.

E. Multi-color fur

The fur is made up of two or three colors. Rarely will you find a fur coat made with these because of the problem of matching the patterns and colors. They are sometimes found in small furred items. Certain color patterns are sought for by judges when showing these types of rabbits.

2.1.3. Classification according to different rabbit sizes

Rabbits under this category are commonly classified by size just like body type and fur into four average weight groups as:

1. **Small -** 2.7 kg (Less than six pounds)
2. **Medium -** 2.7 kg (Over six pounds) to 4.1 kg (9 pounds)
3. **Large -** 4.1 kg (Over 9 pounds) to 5 kg (11 pounds)
4. **Giant -** Over 5 kg (Over 11 pounds)

2.1.4. Classification according to the purpose of rearing

Rabbits under this category are classified according to the purpose they serve. The four main purposes for raising rabbits are:

1. **Fancy**: For show or raised as pets e.g., American Fuzzy lop.
2. **Meat**: Animals that are raised for meat e.g., New Zealand white and Grey giant.
3. **Textile**: Animals raised for their fur e.g., New Zealand white and New Zealand grey.
4. **Research**: Rabbits that are bred and kept specifically for research.
5. **Wool**: Rabbits that are raised for wool production e.g., Angora.

2.1.5. Classification according to rabbit tameness

There are three main rabbit groups. Most of today's domestic breeds have originated from the European Rabbit. These rabbits are classified into the following groups:

A. **Domestic Rabbits** - Rabbits that have been bred by humans for certain traits such as for meat, fur and or size (Table 2.1)

B. **Pet Rabbits** - Domestic rabbits that are specifically raised as pets.

C. **Wild Rabbits** - Rabbits that were born in the wild and currently reside outside of human captivity (Table 2.2.).

Note: Pet Rabbits are basically just a subclass of Domestic Rabbits

Table 2.1 : Domestic rabbits breeds

Breed	Weight (kg)	Size	Body type	Fur
American	4.5 -4.95	Large	Semi Arch	Flyback
American Chinchilla	4.05 - 4.95	Large	Commercial	Rollback
American Fuzzy Lop	1.57 - 1.69	Small	Compact	Wool
American Sable	3.6 - 4.05	Medium	Commercial	Rollback
Belgian Hare	2.7 - 4.28	Medium	Full Arch	Flyback
Beveren	3.8 - 4.95	Large	Semi Arch	Rollback
Blanc de Hotot	3.8 - 4.95	Large	Commercial	Rollback
Britannia Petite	1.12	Small	Full Arch	Flyback
Californian	3.6 - 4.5	Large	Commercial	Flyback
Champagne d'Argent	4.06 - 4.95	Large	Commercial	Flyback
Checkered Giant	5.18+	Giant	Full Arch	Flyback
Cinnamon	3.6 - 4.95	Large	Commercial	Flyback
Creme d'Argent	3.6 - 4.95	Large	Commercial	Flyback
Dutch	1.35 - 2.25	Small	Compact	Flyback
Dwarf Hotot	1.13 - 1.35	Small	Compact	Rollback
English Angora	2.7 - 4.05	Medium	Compact	Wool
English Lop	4.05 - 4.5	Large	Semi Arch	Flyback
English Spot	2.7 - 3.6	Medium	Full Arch	Flyback
Flemish Giant	5.85	Giant	Semi Arch	Rollback
Florida White	2.7	Small	Compact	Flyback
French Angora	2.7 - 4.05	Medium	Commercial	Wool
French Lop	4.73 - 4.95	Large	Commercial	Flyback
Giant Angora	4.06 - 4.95	Large	Commercial	Wool
Gaint Chinchilla	5.4 - 16.75	Giant	Semi Arch	Flyback

Contd.

Harlequin	2.93 - 4.05	Medium	Commercial	Flyback
Havana	2.3 - 2.7	Small	Compact	Flyback
Himalayan	1.35 - 2.25	Small	Cylindrical	Flyback
Breed	Weight (kg)	Size	Body type	Fur
Holland Lop	1.35 - 1.8	Small	Compact	Rollback
Jersey Wooly	1.35 - 1.58	Small	Compact	Wool
Lilac	2.48 - 3.38	Medium	Compact	Rollback
Mini Lop	2.03 - 2.7	Small	Compact	Rollback
Mini Rex	1.35 - 2.03	Small	Compact	Rex
Mini Satin	2.7	Small	Compact	Satin
Netherland Dwarf	0.9 - 1.13	Small	Compact	Rollback
New Zealand	4.06 - 4.95	Large	Commercial	Flyback
Palomino	4.06 - 4.5	Large	Commercial	Flyback
Polish	1.13 - 1.58	Small	Compact	Flyback
Rex	- 4.05	Medium	Commercial	Rex
Rhinelander	2.7 - 4.05	Medium	Full Arch	Flyback
Satin	4.06 - 4.95	Large	Commercial	Satin
Satin Angora	2.7 - 4.05	Medium	Commercial	Wool
Silver	2.7	Small	Compact	Flyback
Silver Fox	4.06 - 4.95	Large	Commercial	Rollback
Silver Marten	2.7 - 4.05	Medium	Commercial	Flyback
Standard Chinchilla	2.25 - 3.15	Medium	Compact	Rollback
Tan	1.8 - 2.7	Small	Full Arch	Flyback
Thrianta	- 4.05	Medium	Compact	Rollback

Size: The rabbit breeds listed above are grouped by size based upon the following standards:

Small: 0.9 - 2.7 kg; Medium: 2.7 - 4kg; Large: 4 - 5 kg; Giant: 7.5 kg +

Table 2.2 : Wild rabbit breeds

Amami Rabbit	**Annamite Rabbit**	**Antelope Jackrabbit**	**Arctic Hare**
Black Jackrabbit	Black-tailed Jackrabbit	Brown Hare	Brush Rabbit
Desert Cottontail	Eastern Cottontail	European Hare	Jack Rabbit
Marsh Rabbit	Mountain Cottontail	Mountain Hare	Pygmy Rabbit
Red Rock Hare	Riparian Brush Rabbit	Riverine Rabbit	Snowshoe Hare
Swamp Rabbit	Sumatra Short-eared Rabbit	Tres Marias Cottontail	Volcano Rabbit
White-sided Jackrabbit	White-tailed Jackrabbit	Wood Rabbits	

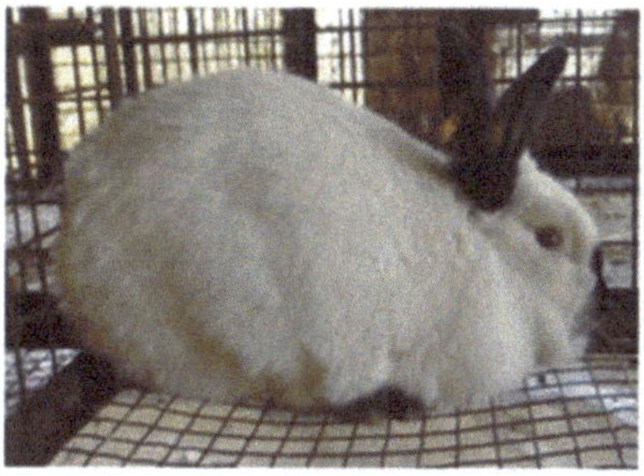

Californian

New Zealand White

Angora

Dutch

Grey Gaint

Himalayan

Chinchilla

Flemish Gaint

Breeds of Rabbit

❑❑❑

Chapter – 3

Housing of Rabbits

The success of rabbit farming depends upon the location and design of the rabbit house. Rabbit is a burrowing animal and has sharp reflexes. It is easily disturbed by any new and sudden noise. The farm should be located in a peaceful environment away from the hustle and bustle of human population. A good housing facility with adequate arrangement of ventilation is the most important factor in rabbit house.

3.1. Ideal Environmental Conditions Needed for Rearing of Rabbits

3.1.1. Light

Exposure to light for eight out of 24 hours favours spermatogenesis and sexual activity in bucks. Conversely, exposure for 14 to 16 hours a day favours female sexual activity and fertilization. Performance is more constant in windowless rabbitries with artificial lighting than in rabbitries which supplement sunlight by artificial lighting. Breeding does need considerable luminosity of at least 30 to 40 lux. Very young rabbits do not really need light, but 15 to 16 hours per day do no harm. Light should not be switched on or off suddenly since this may make the rabbit to leap here and there due to agony leading to fracture and other injuries. A much weaker light (5 to 10 lux) can be used for young rabbits.

3.1.2. Temperature

Rabbits have a constant internal (rectal) temperature so heat production and losses must vary to maintain body temperature. They do this by modifying their feed intake level (regulating production). They use three devices to modify heat loss: general body position, breathing rate and peripheral temperature, especially ear temperature.

If the ambient temperature is low (below 10°C) the animals curl up to minimize the total area losing heat and lower their ear temperature. If the temperature is high (above 25° to 30°C), the animals stretch out so they can lose as much heat as possible by radiation and convection, and step up their ear temperature.

The ears function like a car radiator. The efficiency of the cooling system depends on the air speed around the animal. At the same time the animal pants to increase heat loss through evaporation of water (latent heat). The sweat glands are not functional in rabbits and the only controlled means of latent heat evacuation is by altering the breathing rate. Perspiration (the evacuation of water through skin) is never great because of the fur. These systems work between 0°C and 30°C but when ambient temperatures reach 35°C rabbits can no longer regulate their internal temperature and hyperthermia sets in.

Heat regulation of newborn rabbits is somewhat different: they have no fur and cannot correctly adjust their food intake as the doe's milk output is the result of an involuntary reaction. At birth they have rather good fat reserves which help them maintain body temperature if two conditions are met. The surrounding temperature must be at least 28°C (30° to 32°C if possible), and they must have other young to huddle against to reduce heat loss. At birth, young rabbits cannot modify their body shape by curling up. The only way they can limit heat loss through convection and radiation is to huddle together with the other young in the litter. In fact, if ambient temperature varies during the day the young rabbits will move apart when the temperature is high and huddle back together when it goes down. But a sudden temperature drop may well exhaust their thermoregulation potential before they can get back in the huddle and they can die of cold 10 cm away from the group. The newborn rabbit is blind and the incomplete myelinization of the nervous system that governs motor control hampers coordinated movement. The producer must make sure the temperature in the nest remains constant to prevent this sort of accident.

3.1.3. Humidity

Rabbits are sensitive to very low humidity (below 55 percent) but not to very high humidity. This may be explained by the fact that wild rabbits spend much of their lives in underground burrows with a humidity level near saturation point (100 percent). While the humidity level does not seem to trouble the rabbit in moderate temperatures, this is not so with temperature extremes.

When the temperature is too high (close to the rabbit's body temperature) and humidity is also high, not much latent heat can be exported as water vapour through evaporation. The result is discomfort which can be followed by prostration. Very hot spells with near 100 percent humidity can cause serious problems. This is common in tropical climates during the rainy season.

When the temperature is too low and humidity close to saturation point, water condenses on poorly insulated walls, especially at so-called "heat bridges". Water is a good heat conductor and so the cold becomes more penetrating, causing heat loss in the animals through convection and conduction. Digestive and respiratory disorders often follow. When the surrounding air is cold, excess humidity modifies the secretion and viscosity of the mucus protecting the upper respiratory apparatus. Air which is too dry (below 60 percent relative humidity) and too hot is even more dangerous. Not only does it upset the secretion of mucus but the ensuing evaporation shrinks the size of the droplets carrying infection agents, enabling them to penetrate more easily the respiratory apparatus.

3.1.4. Ventillation

The rabbitry must have enough ventilation to evacuate the harmful gases given off by the rabbits (CO_2), to renew the oxygen and get rid of excess humidity (evaporation, exhalation) and excess heat given off by the rabbits. Ventilation needs can vary enormously, depending especially on climate, cage type and population density. It is relatively easy and cheap to measure temperature and humidity, but exact air flow measurement requires sophisticated, expensive, hard-to-get equipment such as a hot-wire anemometer (a revolving-cup anemometer is not sensitive enough). However, the producer can estimate the rate of air flow near rabbits by using a candle flame (Fig. 3.1).

High ammonia air levels, 20 to 30 parts per million (ppm), greatly weaken the rabbit's upper respiratory tract and open the door to bacteria such as pasteurella and bordetella. To keep NH_3 levels down, ventilation can be increased. The risk is then over ventilation, with all the negative consequences (Fig. 3.2). A more effective solution is to limit NH_3 production from fermenting floor litter (droppings and urine) by removing the litter quickly or keeping it dry. The maximum permissible NH_3 content in the air rabbits breathe is 5 ppm.

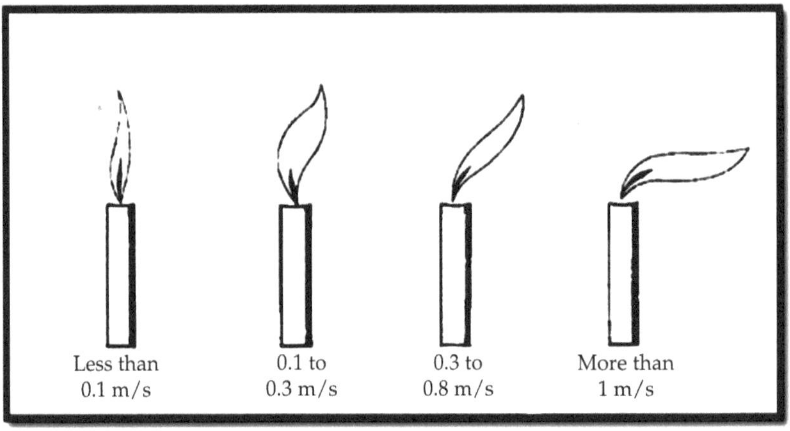

Fig. 3.1 : Estimating air flow with a candle flame

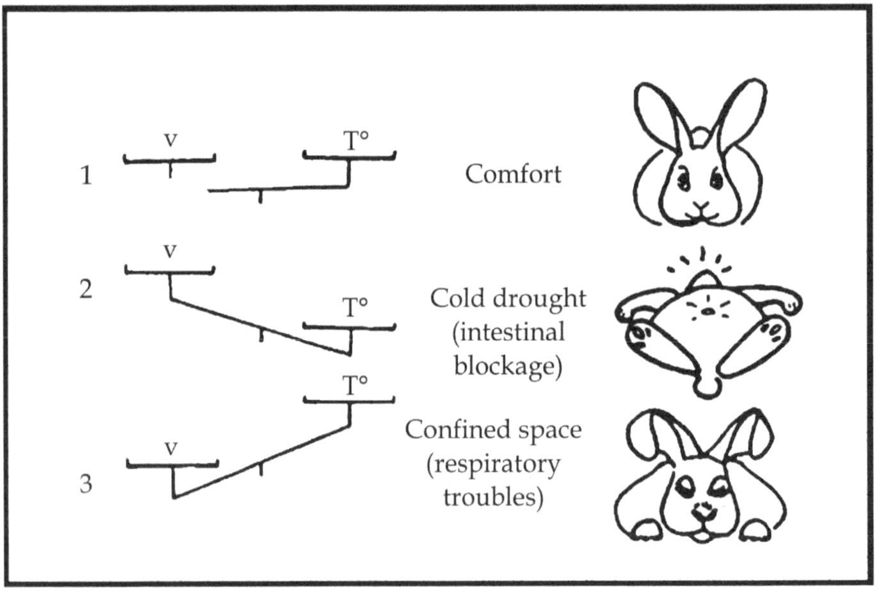

Fig. 3.2 : Effect of air speed (V) and temperature (T°) on health of rabbits

3.2. Housing Systems

3.2.1. Rabbit house

An ideal commercial rabbit house should be constructed with modern housing equipments. Attention should be given to maintain proper temperature, ventilation, humidity and lighting aspects of the proposed house. A shed with 4.2 m central height, 3.0 m side height may be ideal.

3.2.1.1. Floor

The floor should be designed in such a way so that automatic washing and drainage device can be operated. Preferably it should be made up of concrete since it has an advantage of preventing rodents to make holes and nests and also can be cleaned and disinfected easily. If mud floors are made then they should be made hard through ramming to prevent hole making and digging by rodents in the farm. The floor should by connected to drainage pocket and should have enough slope to allow drainage of water.

3.2.1.2. Roof

The roof must have sloping gradient and can be made of various materials like corrugated tin or asbestos. Thatched roofs can also be used but it requires changing after some time. Straw or palm leaves can be put on the roof to reduce heat during summer. Roof can be raised on polls made of bamboo, concrete pillars or iron beam.

3.2.1.3. Walls

The walls should be preferably made of bricks since it has got the advantage of being durable and providing protection. Bamboo mats along with polythene sheets can also be used as walls in backyard rabbitry. Some cheap material like gunny bags or feed bags can also be used to reduce housing cost. The open spaces of walls should be covered with wire nets. Size of the mesh should be selected in such a way so that predators like rats, mice, snakes etc cannot enter.

3.2.1.4. Doors

Doors can be made of bamboo mat, wires, wooden plank or tin sheet.

The above requirements can be met by either outdoor or indoor hutches.

3.2.2. Cage system

Rabbit cages (Hutches) can be purchased or locally constructed. These cages are generally made of wood, RCC sheets, asbestos sheets, welded wire, asphaltic or tin sheets. There are two types of hutches adopted by most of the farmers based on the number of rabbits to be housed.

a) Individual cages

Individual cage type of housing is adopted on a small scale and backyard rabbit farming. Under this system no large shed with suspended cages is needed and the rabbits are housed in individual cages which has a solid roof to provide shelter from rain and direct sunrays.

More detailed recommendations are given below:

Floor Space:	**Small scale forage system:** 1 sq. m. per adult rabbit. **Large scale intensive system:** 0-4 kg live weight - 0.35 sq. m. per rabbit. 4-6 kg live weight - 0.50 sq. m. per rabbit. 0.2 sq. m. extra for the litter.
Floor Design:	Floor should slope outwards if solid wire floors should have 1 -1.5 cm square and made of galvanized wire slats should be 2.5 cm wide with 1 cm gap.
Height:	Tall enough to allow the rabbit to stretch fully (50- 70 cm).

Cages can be kept either outdoor or indoor depending upon the environmental conditions and availability of hand cages.

- **Outdoor cages:** These cages are kept outside all the time.
- **Indoor cages:** These cages are kept inside a house or open shed.

The dimension of a typical cage for general purpose are as follows:

Height from ground	Approximately 1 m
Height of cage	60 cm at front; 50 cm at the back
Width	50 - 60 cm
Length	90 - 120 cm

The dimension of various cages is given below:

	Length	Breadth	Height
Breeding doe with kits	90 cm	60 cm	60 cm
Non - breeding doe	60-75 cm	60 cm	60 cm
Buck	60-75 cm	60 cm	60 cm
Colony cage	90 cm	60 cm	60 cm

b) Colony cages

Colony cages are of the same dimensions as for breeding doe cage. The dimensions can be increased with length and breadth as the number of weaners is increased.

Hutches can be used with or without litter. As per this, two types of cages can be used for rearing of rabbits as follows:

1. Cages (hutches) with litter

Traditionally straw litter is used but this material can be replaced by any other dry fibrous product which is not rough to the touch (soft shavings, hay, industrial cotton waste and so on). The cages are either of concrete (lasting 15 to 30 years) or wood (lasting not more than two years). Cages for breeding animals usually have at least a 60 to 70 cm x 80 to 100 cm floor space and are 50 to 60 cm high. Identical cages are often used for fattening five or six young rabbits (to 2.5 to 2.8 kg). The litter should be replaced weekly to control parasitism.

A variation called "Deep litter" is used in slightly taller cages. The floor is covered with a bed (minimum thickness 15 to 20 cm) of absorbent material (turf, wood shavings) evenly covered with straw. Every six or seven weeks the whole lot, absorbing layer plus accumulated straw, has to be replaced. This system saves labour in cleaning and has the advantage of the comfort provided by the straw, but it does use a great deal of absorbent material so it is only applicable where this material is both readily available and cheap.

2. Cages (hutches) without litter

Rabbits are raised above the ground on a litterless floor (Wire-mesh or slatted floor). Wire-mesh floors should be thick enough not to injure the pads of the rabbits' feet (diameter 2.4 mm, minimum 2 mm); the mesh should be wide enough to let the droppings fall through (diameter 1 to 1.3 cm, according to feed) but narrow enough to prevent the feet getting caught in the mesh.

Good commercial meshes (25 x 13 mm, 76 x 13 mm or 19 x 19 mm) are available. To avoid injury to the rabbits' feet the wire is welded and then galvanized. Plastic mesh is impractical because no plastic material can withstand the animals' gnawing.

Various kinds of slats like slats of wood, bamboo, plastic and metal can be used but the individual slats of the structure have to be spaced about 1.3 to 1.5 cm apart so droppings can fall through. Wherever possible, wire mesh is preferable to slats. If slats are used instead of mesh, bamboo should be preferred to wood if possible.

Only lightweight, calm animals or specially selected breeds (New Zealand White, Califorian) can be raised entirely on wiremesh floors. Producers often compromise by raising the male and female breeding animals on litter and the fattening stock on mesh floors. The breeding animals of heavy breeds can be reared on slatted floors and the young on wire mesh; but slatted floors must be cleaned more frequently.

As per the design, cages may of the following types:

1. Flat-deck system (Fig 3.3)
2. Californian system (Fig 3.4)
3. Inclined-slope battery system (Fig 3.5)
4. Compact battery system (Fig 3.6)

1. Flat-deck system

Flat-deck system is preferred because it has the lowest rabbit density, thus alleviating the problems which arise when many animals are reared in the same building (Fig. 3.3). In this system the cages are all on one level. They usually open at the top. They can be suspended by chains or set on feet or low walls. Floor litter drops into pits (ranging in depth from 20 cm to 1.5 m). Shallow pits are cleared daily or every two or three days and deep pits every one to three years.

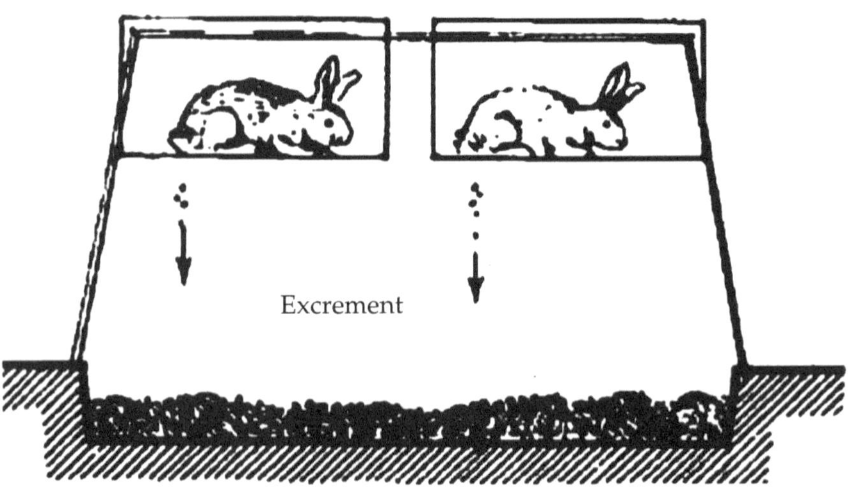

Fig. 3.3 : Flat deck cage

Advantages:

- Convenient supervision and handling of animals.
- Long life for the material used.
- Animal and producer comfort.
- No elaborate ventilation system required.

Drawbacks

Low animal concentration per square metre of building, which increases the investment per animal housed despite the low-cost cages. However, flat-decks are recommended for nurseries. They could also be used for fattening, but at a higher investment cost per cage.

2. California cage system

In this system the cages are staggered, one deck higher than the other but not above it (Fig. 3.4). The cages on the lower level open at the top and those on the upper level at the front (poorer access). Floor litter drops beneath the cages and is collected as in the flat-deck system.

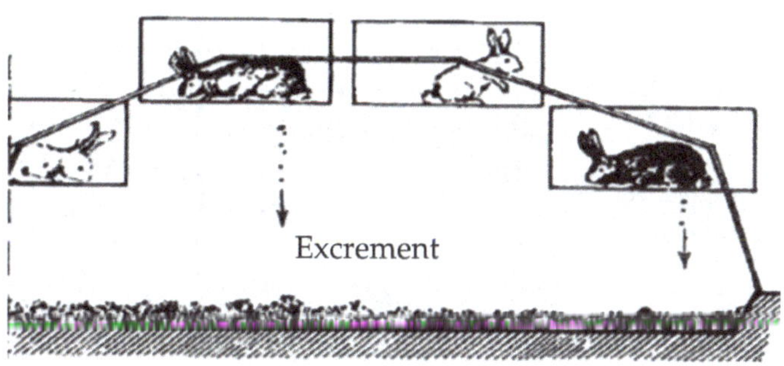

Fig. 3.4: Californian cage

Advantages:

Same advantages as the flat-deck with regard to ventilation; slight increase in animal density per square metre of building.

Drawbacks

- Access to upper cages and supervision difficult
- Frame more expensive than flat-deck.

3. Inclined-slope battery system

The cages are placed one above the other (Fig. 3.5). Waste slides down asbestos or metal panels into troughs from which it is removed manually with scrapers or with rurming water. Cages obviously open at the front.

Advantages:

- Higher animal density
- Reasonable cost although more expensive than flat-deck.

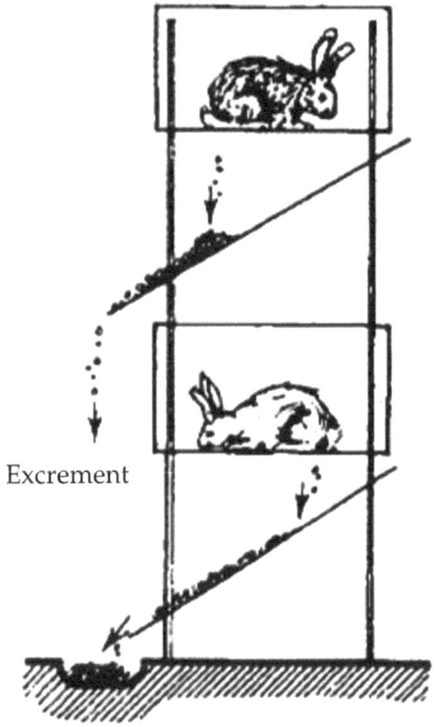

Fig. 3.5 : Inclined-slope battery cage

Drawbacks

- No matter what material is used for the panels or how steeply they slant, waste does not drop properly and must be periodically raked down
- High animal density demands careful ventilation
- No access to the cages, supervision and handling of the animals is more difficult.

4. Compact batteries

Waste can be removed by conveyor belt or vats can be installed beneath the cages and emptied by cable operated scrapers (manual or electric). As with the inclined-slope battery, the compact battery cages (Fig. 3.6) must open from the front.

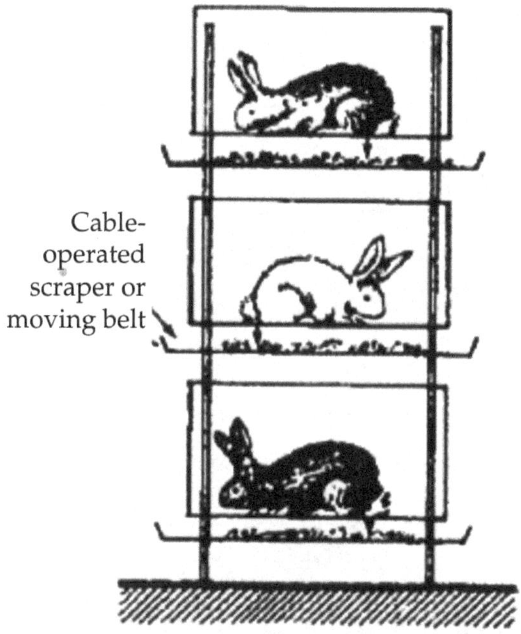

Fig. 3.6 : Compact battery cage

Advantage:

The advantage with this system is that the maximum density of animals reduces costs per animal housed.

Drawbacks

- As for the inclined-slope battery regarding ventilation, access to cages, supervision and handling of animals is difficult.
- Quicker wear and tear on materials.
- With automatic scraping there is the risk of breakdown and harmful gases from the scrapers.
- Poor distribution of light for breeding does.

Fig 3.6.1 : Rabbit Cages (Battery system)

Fig. 3.6.2 : Interior view of an individual cages

Fig. 3.6.3 : Exterior view of individual cages

3.3. Equipments Required in Cages

3.3.1. Watering equipment

A permanent dispenser of clean water is an essential item in each cage, wherever rabbits are not fed green forage alone. Using old cans or glass or earthenware pots as drinkers can create a hygiene problem. Rabbits tend to soil their water, especially if they are reared on straw litter. The drinkers should be fastened so that the rabbits cannot overturn them and so that the breeder can easily clean and refill them once or twice a day. Drinkers can be of two types:

1. Inverted water bottle drinker
2. Automatic drinker

1. Inverted water bottle drinker

A bottle is inverted over a small trough, which is small enough to restrict pollution (Fig. 3.7). The bottle is big so that it needs to be refilled less often and the breeder can see at a glance whether the animals' water intake is normal.

2. Automatic drinker

Automatic drinkers (Fig. 3.7) are fed by water from a low-pressure tank 50 to 150 cm above cage level. This tank can be used to administer medicine with the water. It is usually filled either by water under pressure (automatic watering) or manually (semi-automatic watering). The tank must be in the shade so that the water will not heat, which would be bad for the rabbits.

Automatic drinkers are of 2 types:

a. Open automatic drinker

b. Nipple drinker

a. Open automatic drinker

The open drinker guarantees that the rabbits will be watered but it is expensive and there is a high risk of water pollution.

b. Nipple drinker

A nipple drinker requires some learning on the part of the rabbits and wastes water. Even if there is no leak, the rabbits do not drink all the water that drips out. This can then wet litter or waste. The cost, however, is usually half that of an automatic open drinker. Above all, it ensures that the rabbits will always have clean water. A nipple drinker is the only kind that can be used if the rabbits are fed meal.

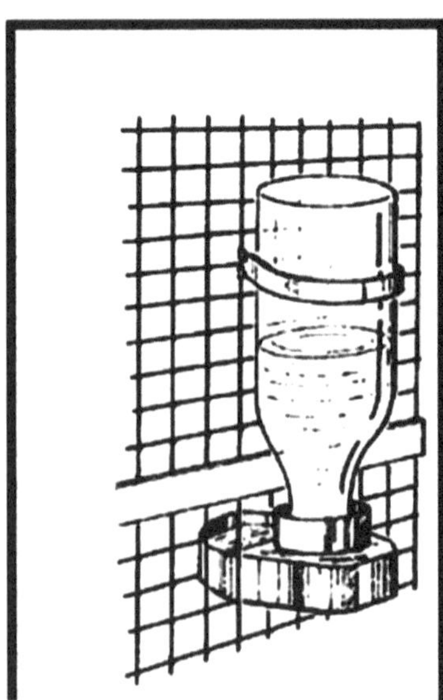

Fig. 3.7 : Inverted water bottle drinker

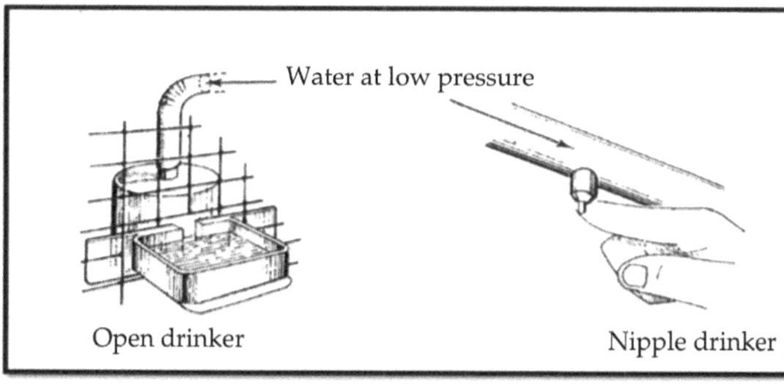

Fig. 3.8 : Automatic drinkers

3.3.2. Feed troughs and racks

Cages should be fitted with troughs or (feed hoppers shown in Fig.3.9 for grain or pellets, small troughs for feed mashes) or forage racks, or both, depending on how the rabbits are to be fed. Troughs must be easy to clean and disinfect, so they should be detachable. Troughs and racks should be easy to fill from outside the cage without having to open the access door, but the feed must be protected from bad weather and predators. The racks should hold at least one day's ration, the hoppers enough for two or three days, and the mash troughs a single ration. The bars of the rack should be strong enough to withstand the rabbits' teeth and keep out the young rabbits, who like to lie on the forage but soil it. The feed hopper should also have a trap to keep the young out. The width between partitions in the feed box should be about 7 to 8 cm for medium breeds. The bars of the racks can be more closely spaced (1 to 2 cm) to prevent waste.

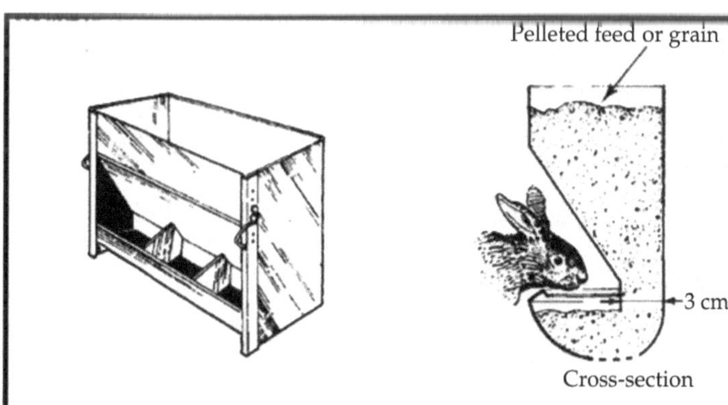

Fig. 3.9 : Feeding hopper

3.3.3. Nest box

The nest box (Fig. 3.10 & 3.11) should be considered one of the most important items of equipment in rabbit production. It directly affects the viability of the young in the preweaning stage, which is the high-risk mortality period (15 to 40 percent of liveborn rabbits). The job of the nest box is to reproduce conditions in the burrow of a wild doe and protect the young against attacks from the outside environment so that they can get through the first few difficult days of life in optimal comfort. To do this the nest box must:

- Allow the doe to kindle and nurse her young in comfort.
- Keep the young in a healthy, clean environment.
- Prevent dampness from the animals' urine.
- Keep the young together in cold weather and help them maintain a constant temperature close to 30° to 35°C in the middle of the nest.
- In hot weather, allow the doe to scatter the nest so that the young can adapt to the temperature.
- Keep the young from leaving the nest too early and make it easy for them to get back if they do get out.
- Allow the producer to monitor the litter, remove any dead animals, introduce baby rabbits to be fostered and change bedding material easily, without disturbing the doe and the young.

The nest box is strongly advised for rabbitries using straw litter; it is essential in modern production. The box most recommended to meet these requirements, especially the doe's comfort when kindling and nursing, is a rectangular paralleled pipe at least 50 x 25 x 25 cm. If there is a dividing panel to keep the young together, at least 30 x 30 cm must be left on that side so that the mother can nurse them in comfort.

The nest box should be made from materials that are impossible to gnaw, easy to disinfect, insulating and resistant to moisture. In a well-heated rabbitry or a warm climate, galvanized iron can be used if some other material such as plywood, wood or plastic is used for the bottom.

To comply with the habits of the doe and her young, and to make the breeder's work easier, the box should have the following features:

- The bottom should be hollowed so that the young can huddle together when the temperature drops, but it should also favour their dispersion when the temperature rises.

- The bottom should be non-skid, to prevent dislocation of the young bones ("Swimming").
- Access for the doe should be opposite the section holding the young so that she will not crush them when entering the nest box suddenly.
- The access hole for the dam should be fairly narrow, square or round, and about 15 cm across.
- The bottom of the box should be designed to allow urine to run off. It can be perforated or a space 1 to 1.5 cm wide can be left between the floor and the sides of the box. Another alternative is straw sandwiched between two layers of mesh.
- The bottom should be detachable so that the whole interior of the box can be cleaned.
- The top should have a trapdoor so that the breeder can easily observe and check the rabbits.
- There should be a sufficiently high ledge, level with the doe's access hole, to keep the young from leaving the box too early (before day 15). An even better solution is to install the box below the level of the cage floors so that the babies can get back easily.

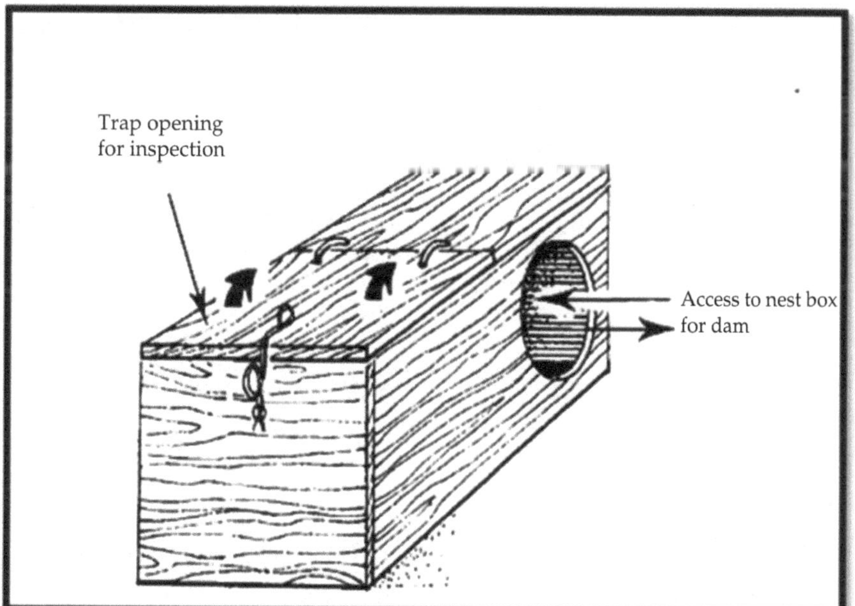

Fig. 3.10 : Wooden nest box

Fig. 3.11 : Design of a nest box

The doe needs materials in addition to her own fur to make a good nest. Clean straw or soft, untreated wood chips are suitable and dried grasses can be used. Cellulose cottonwool must never be used. The nest box can be placed inside or outside the cage. If it is outside it can be fastened to the side of the cage or preferably to the front, to make inspection easier.

❑❑❑

Chapter – 4

Nutrition of Rabbit

4.1. Digestive System of Rabbit

The digestive system of rabbit includes:

A. The digestive tract

B. Its associated glands

The digestive system breaks down the complex molecules into the simpler one so that it is readily absorbed from gastro intestinal mucosa.

A. The digestive tract (gastro intestinal tract/ alimentary canal)

The digestive tract is essentially a tubular structure for the ingestion and digestion of food. In an adult (4 to 4.5 kg) or semi-adult (2.5 to 3 kg) rabbit the total length of the alimentary canal is 4.5 to 5 m. It commences from mouth and terminates in anus. G.I.T is divided into following parts:

1. Mouth

The mouth cavity (Oral/buccal cavity) of rabbit is oval and relatively small. It is bounded by movable lips. The upper lip has a small cleft in the middle from where the incisor teeth are exposed. The roof of the buccal cavity is called palate. The palate is divided into the anterior hard and posterior soft palate. The palate separates the buccal cavity from the nasal cavity.

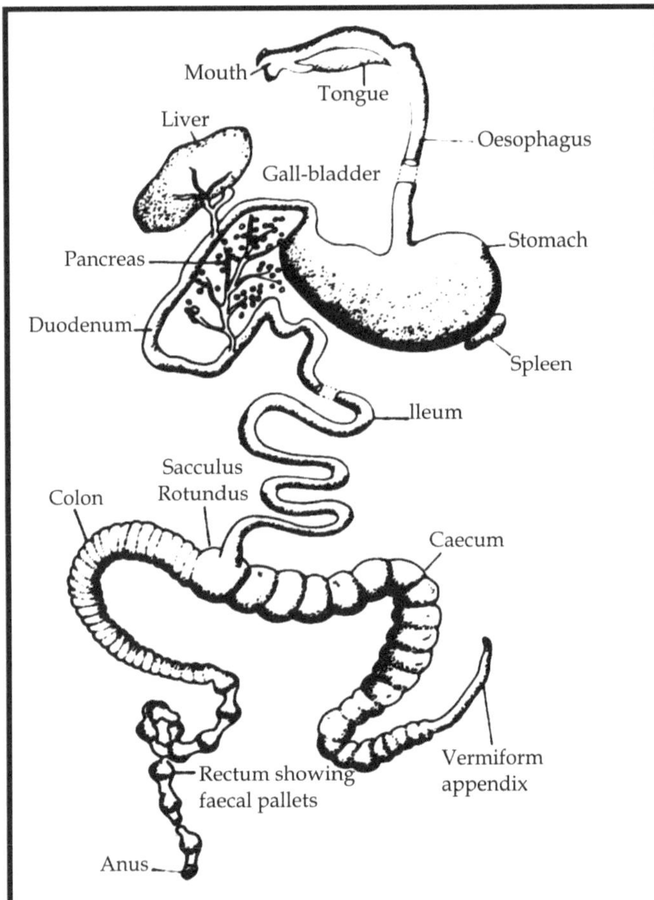

Fig. 4.1: Digestive system of rabbit

2. Tongue

The tongue is muscular. It protrudes slightly along the floor of the oral cavity. The tongue plays an important role in moving food from the incisors back to the cheek teeth where it is ground.

3. Teeth

An adult rabbit has 28 permanent teeth. Rabbits and other lagomorphs usually shed their deciduous teeth before (or very shortly after) their birth, and are usually born with their permanent teeth. Rabbits have **aradicular hypsodont teeth** (including the incisors, pre molars and molars) which continue to erupt and grow through out life.This is sometimes referred to as an **elodent dentition.** They grow on average 8-12 cm per year (1cm per month) and do not have true roots. Rodents unlike rabbits have permanently growing incisors only. Rabbit dentition includes incisors and cheek teeth(Molars and premolars) where as canine teeth are absent.

a. Incisors

Rabbits have six incisors (Three pairs):

Two pairs of incisors in the upper jaw or maxilla (the second pair, often referred to as "Peg teeth" are much smaller and are positioned behind the longer front incisors)and one pair of lower incisors on the lower jaw or mandible.The additional set of upper incisors (peg teeth) separate them anatomically from rodents.

b. Cheek teeth

Molars and premolars are together called Cheek teeth. Rabbits have three upper premolars, three upper molars, two lower premolars, and two lower molars on each side. There are more "cheek teeth" on the maxilla than on the mandible, and the mandible is narrower than the maxilla. The cheek teeth are separated from the incisors by a fairly long space of cleft (diastema), which is extremely helpful when syringe feeding medicines to a rabbit.

Three to four millimeters of incisors are worn away every week, whereas the posterior teeth require a month to wear away the same amount. The four larger incisor teeth are used to grasp, tear, and slice, while the "cheek teeth" are used to grind and chomp. When at rest, the tips of the lower incisors are normally positioned between the maxillary first incisors and the peg teeth. The cheek teeth, however, should be slightly out of alignment in the "at rest" position. The cheek teeth are brought into alignment (occlusion) when the mandible (jaw) is slightly retracted and angled so that the jaw joint (temporomandibular joint) can reposition itself, allowing the cheek teeth to come together, while at the same time, separating the incisors. This allows the cheek teeth to move laterally while chewing, without interference from the incisors.

Dental Formula: [i 2/1, c 0/0, pm 3/2, m 3/3] x 2 = 28

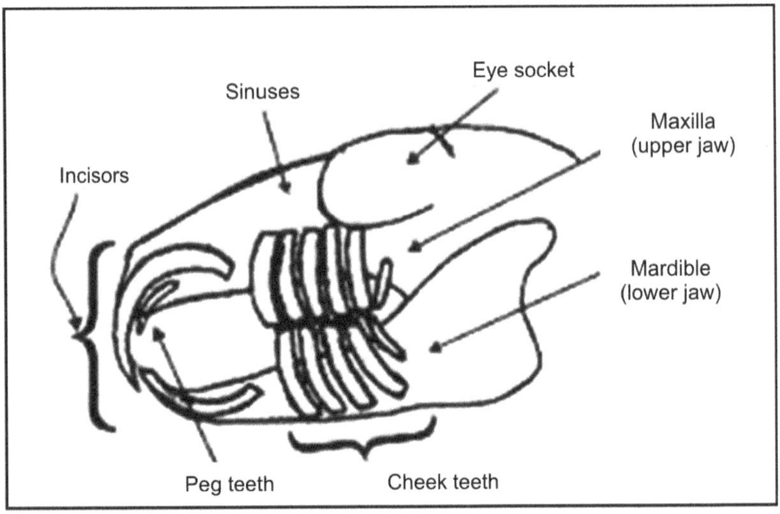

Fig. 4.2: Skull of rabbit

Potential dental conditions

- Elongation of cheek tooth crown
- Spurs on the cheek teeth
- Decay of bone or teeth (caries)
- Soft tissue lesions, often located on the sides of the tongue, underneath the tongue, or on the cheeks
- Root elongation of the cheek teeth
- Swelling along the mandible (frequently mandible abscesses will develop which can lead to osteomyelitis)
- Bulging of the eye (sometimes a retrobulbar abscess will develop as a result of dental problems)
- Ocular discharge (sometimes caused by a tooth root obstructing the nasolacrimal duct)
- Food impaction around cheek teeth which can lead to infection
- Periodontal abscesses or infection
- Incisor malocclusion
- Temporomandibular joint problems (TMJ)
- Fractures of incisor roots

4. Pharynx

The oral cavity leads to pharynx. The dorsal part is known as nasopharynx and the ventral part as oropharynx. Oropharynx leads to the oesophagus and nasopharynx leads to larynx. The entrance of larynx and the glottis, is guarded by the epiglottis. The Eustachian tubes from the middle ear also opens into the pharynx.

5. Oesophagus

It is a long cylindrical elastic tube which enters the abdomen through an aperture in the diaphragm and opens into the stomach.

6. Stomach

After oesophagus there is a simple stomach. It is a large **pear shaped** sac which stores about 90 to 100 g of a rather pasty mixture of feedstuffs. There is a valvular arrangement called cardia to prevent the regurgitation of food. This region is known as cardiac end of the stomach. There are gastric glands which produce digestive juices.

7. Small intestine

Small intestine is about 3 m long and 0.8 to 1 cm in diameter. The pyloric part of the stomach leads to the first part of the intestine known as duodenum. Duodenum takes a U shaped loop in the mesentry on which the pancreas lies. The bile duct opens into the duodenum and the pancreatic duct opens into the distal limb of the duodenum. Small intestine is the largest part of alimentary canal. The coiled part of the small intestine is known as ileum. The mucous membrane of the small intestine is folded to form finger like projections called villi. Each villi is supplied with blood capillaries and lymphatics. Numerous Brunner's gland open into the crypts (The depression between the villi). The secretion of the crypts of Lieberkuhn and Brunner's glands form the intestinal juice known as Succus entericus. At the junction of the ileum with the large intestine, there is blind ending diverticulum called caecum. Caecum is nearly 46 cm in length and about 25mm in width. It contains 100 to 120 g of a uniform pasty rnix with a dry matter content of about 22 percent. It ends into a narrow vermiform appendix, which measures about 10 cm in length and has a much smaller diameter at the end. Its walls are composed of lymph tissues.This three way junction is marked by a swelling, the sacculus rotundus into which the ileum opens and with which there is a valvular arrangement the ilio-caecal value. In rabbit, which is purely vegetarian the caecum consists of a series of pouch like dilations. The vermiform appendix has a smooth surface.

8. Large intestine

It has three parts known as colon, rectum and anus. Colon is sacculated and about 80 cm in length. Rectum is about 45 cm in length. Rectum opens exteriorly at the anus.

B. Digestive glands

The various digestive glands associated with the alimentary canal are:

1. Salivary gland
2. Liver
3. Pancreas
4. Gastric glands of stomach and intestine.

Liver and pancreas are the major glands that secrete into the small intestin. Bile from the liver contains bile salts and many organic substances which aid digestion, but has no enzymes. The reverse is true of pancreatic juice which contains a sizeable quantity of digestive enzymes allowing the breakdown of proteins (trypsin, chymotrypsin), starch (amylase) and fats (lipase).

4.2. Mechanism of Digestion

The mechanism of digestion is summarized in the Table 4.1

Table 4.1: Mechanism of digestion

Region	Juice	Enzyme	Substance	End Product
Mouth	Saliva	Ptylin	Starch and glycogen	Maltose
Stomach	Gastric Juice	Pepsin + HCl	Protein	Peptones & roteoses
		Rennin	Caseinogen	Casein
		Gastric lipase	Emulsified	Fatty acid & glycerol
Duodenum	Bile		Sodium carbonate	Neutralizes acidity of chyme
			Sodium glycocholate Sodium taurocholate	(a) Stimulate pancreatic lipase (b) Reduce surface tension of fat
Pancreas	Pancreatic Juice	Amylase Lipase	StarchFat	Maltose Glycerol & Fatty Acid

4.3. Digestive Tract and Caecatrophy

Feed eaten by the rabbit quickly reaches the stomach. There it finds an acid environment. It remains in the stomach for a few hours (three to six), undergoing little chemical change. The contents of the stomach are gradually "injected" into the small intestine in short bursts, by strong stomach contractions. As the contents enter the small intestine they are diluted by the flow of bile, the first intestinal secretions and finally the pancreatic juice.

After enzymatic action from these last two secretions the elements that can easily be broken down are freed and pass through the intestinal wall to be carried by the blood to the cells. The particles that are not broken down after a total stay of about one and a half hours in the small intestine enter the caecum. There they have to stay for a certain time, from two to 12 hours, while they are attacked by bacterial enzymes. Elements which can be broken down by this new attack (mainly volatile fatty acids) are freed and in turn pass through the wall of the digestive tract and into the bloodstream.

The contents of the caecum are then evacuated into the colon. Approximately half consists of both large and small food particles not already broken down, while the other half consists of bacteria that have developed in the caecum, fed on matter from the small intestine.

So far, the functioning of the rabbit's digestive tract is virtually the same as that of other monogastric animals. Its uniqueness lies in the dual function of the proximal colon. If the caecum contents enter the colon in the early part of the morning they undergo few biochemical changes. The colon wall secretes a mucus which gradually envelops the pellets formed by the wall contractions. These pellets gather in elongated clusters and are called Soft or Night Pellets (more scientifically, caecotrophes).If the caecal contents enter the colon at another time of day the reaction of the proximal colon is entirely different.

Successive waves of contractions in alternating directions begin to act; the first to evacuate the contents normally and the second to push them back into the caecum. Under the varying pressure and rhythm of these contractions the contents are squeezed like a sponge. Most of the liquid part, containing soluble products and small partices of less than 0.1 mm, is forced back into the caecum. The solid part, containing mainly large particles over 0.3 mm long, forms hard pellets which are then expelled. In fact, as a result of this dual action, the colon produces two types of excrement: hard and soft.

The hard pellets are expelled, but the soft pellets are recovered by the rabbit directly upon being expelled from the anus. To do this the rabbit twists itself round, sucks in the soft faeces as they emerge from the anus, then swallows without chewing them. The rabbit can retrieve the soft pellets easily, even from a mesh floor. By the end of the morning there are large numbers of these pellets inside the stomach, where they may comprise three quarters of the total contents.

From then on the soft pellets follow the same digestive process as normal feed. Considering the fact that some parts of the intake may be recycled once, twice and even three or four times, and depending on the type of feed, the rabbit's digestive process lasts from 18 to 30 hours in all, averaging 20 hours.

The soft pellets consist half of imperfectly broken-down food residues and what is left of the gastric secretions and half of bacteria. The latter contain an appreciable amount of high-value proteins and water soluble vitamins. The practice of caecotrophy therefore has a certain nutritional value.

The composition (Table 4.2) of the soft pellets and the quantity expelled daily are relatively independent of the type of feed ingested, since the bacteria remain constant. In particular, the amount of dry matter recycled daily through caecotrophy is independent of the fibre content of the feed. The higher the crude content of the feed and/or the coarser the particles, the sooner it passes through the digestive tract.

Table 4.2: Composition of hard and soft faeces: averages and range for ten different feeds

Components	Hard pellets		Soft pellets	
	Average	Range	Average	Range
		(Percentage)		
Moisture	41.7	34-52	72.9	63-82
Dry matter	58.3	48-66	27.1	18-37
		(Percentage of dry matter)		
Proteins	13.1	9-25	29.5	21-37
Crude fibre	37.8	22-54	22.0	14-33
Fats	2.6	1.3-5.3	2.4	1.0-4.6
Minerals	8.9	3.1-14.4	10.8	6.4-10.8
Nitrogen-free extract	37.7	28-49	35.1	29-43

On the other hand, this particular function requires roughage. If the feed contains few large particles and / or it is highly digestible, most of the caecal contents are pushed back to the caecum and lose elements which nourish the "normal" bacteria living in the caecum. This would appear to increase the risk of undesirable bacteria developing in this impoverished environment, some of which might be harmful.

It is thus advisable to include a minimum of roughage in the feed, enabling the rabbit's digestive process to be completed fairly rapidly. In theory, roughage is provided by the crude-fibre content of the feed, as this is normally rather hard to digest. However, certain fibre sources (beetroot pulp, fruit pulp in general) are highly digestible (digestibility of crude fibre varies from 60 to 80 percent). Recommendations now made on quantities of indigestible crude fibre to be fed are therefore given below.

Caecotrophy regulation depends on the integrity of the digestive flora and is governed by intake rate. Experiments have shown that caecotrophy starts eight to 12 hours after the feeding of rationed animals, or after the intake peak of animals fed ad lib. In the latter case, the intake rate and hence the function of caecotrophy are governed by the light regime to which the animals are subjected.

Caecotrophy also depends on internal regulatory processes as yet not understood. In particular, the removal of the adrenals halts caecotrophy. Cortisone injections of animals without adrenals causes the resurnption of normal behaviour. The digestive process of the rabbit appears to be highly dependent on adrenalin secretions. Hypersecretion associated with stress slows down digestive activity and entails a high risk of digestive ailments.

Caecotrophy first starts to function in young rabbits (domesticated or wild) at the age of about three weeks, when they start eating solid feed in addition to mother's milk.

4.4. Feeding Behaviour

Feeding behaviour studies have basically involved rabbits receiving balanced concentrates or fed ad lib on dry feed (cereals, straw, dry forage). The feeding pattern of newborn rabbits is imposed by the dam. A doe feeds her young only once every 24 hours (although some does will nurse their young twice). Suckling lasts only two or three minutes. If there is not enough milk the young try to feed every time the doe enters the nestbox, but she will hold back her milk. This behaviour signals insufficient milk production in the doe.

From the third week of life the young rabbits begin to move about, taking a few grams of mother's milk and a little drinking water if available. In a few days the intake of solid feed and water will exceed the milk intake. During this period the changes in feeding behaviour are remarkable: the young rabbit goes from a single milk feed a day to a large number of alternating solid and liquid feeds distributed irregularly throughout the day: 25 to 30 solid or liquid meals every 24 hours.

The number of solid meals, stable up to 12 weeks, tends to decrease slightly thereafter. The total feeding time in a 24-hour period exceeds three hours at age six weeks. It then drops off rapidly, to less than two hours. At any age, feed containing over 70 percent water, such as green forage, will provide rabbits with ample water at temperatures below 20°C.

The consumption of solid and liquid intake fluctuates over a 24-hour period. Much more liquid and solid feed is consumed in the dark than in the light.

Intake in experimental hutches is very high just before the lights are switched off. As the rabbit grows older the nocturnal nature of its feeding habits becomes more pronounced. The number of feeds during light periods drops and the morning "feeding rest" tends to lengthen. The feeding habits of wild rabbits are even more nocturnal than those of domesticated rabbits.

The intake of feed and water depends on the kind of feed and also on the type of rabbit and its age and stage of production. A doe's ad lib feeding during the reproduction cycle varies greatly.The intake during the final days of pregnancy drops off markedly. Some does refuse solid food just before kindling. Water intake, however, never stops completely. After kindling, the feed intake increases very rapidly and can exceed 100 g dry matter / kg live weight a day. Water intake is also high at that time: from 200 to 250 g a day per kg of live weight. A doe that is both pregnant and lactating will eat the same amount as a doe that is lactating only.

4.5. Principle Components of the Rabbit Diet

The principal components of the rabbit diet are fibre, protein, carbohydrate, vitamins and minerals. Of these fibre is the most important.

4.5.1. Fibre

Fibre is the single most important component of the rabbit diet. The particle size and digestibility of the fibre are also important. The rabbit needs large indigestible fibre particles (containing lignocellulose) to drive the gastrointestinal system, and to maintain healthy peristalsis. Gastric stasis and the subsequent accumulation of a 'felt' of ingesta and hair is rarely seen in rabbits on high fibre diets. Large fibre particles ensure the natural wear of the molar teeth with a side-to-side chewing action. Small particles tend to be eaten with a crushing action which encourages unnatural wear on the teeth. High fibre, low energy diets will result in complete caecotrophy, and avoid the development of 'sticky bottom syndrome'. Such diets also prevent obesity. High fibre (hay) diets allow the rabbit to 'graze' for several hours daily, which closely parallels the behaviour of wild rabbits, and reduces boredom and stress. This reduces fur chewing of itself or its companions, and reduces destructive chewing of the environment. Fibre should make up 18±20% of the diet, and of this at least 10% should be crude indigestible fibre. As most complementary dry foods contain less than this (between 5.7 and 14%) every diet should be complemented with good hay. Fibre is particularly important for the maintenance of a healthy digestive system: a high fibre diet keeps the correct balance of bacterial flora in the caecum. If the fibre content of the diet falls, the caecal pH alters, and the populations of Clostridia and Escherichia coli rise. This is particularly significant at lifestages when the gastrointestinal system is most vulnerable, such as weaning, or during antibiotic usage. Rabbits on high fibre diets rarely suffer from mucoid enteritis, or enterotoxaemia. High fibre diets will also protect against enterotoxaemia when antibiotics are given.

4.5.2. Protein

The recommended protein level of the diet is 12±13%. Higher protein levels of 17±18% may be used for breeding stock, and 15% for growth. Protein is part of the concentrated ration, but is also found in hay. High protein diets will lead to an increase in the production and excretion of ammonia, and this may increase the rabbit's susceptibility to respiratory disorders, and eye infections. Excess protein may also lead to incomplete coprophagy and the development of 'Sticky Bottom Syndrome'.

4.5.3. Fat

The dietary requirement for fat is 1±3%; 1% fat is suitable for maintenance, and 3% for pregnancy and growth. However, many rations have increased fat content to increase palatability. The fat content of concentrated food is listed in the food analysis as oil. Vegetable oils, soya and linseed are more digestible than animal fats. Excess dietary fat may lead to the development of arteriosclerosis. A fat deficiency is rare; however, the oils in the dry food are the rabbit's only source of essential fatty acids, and if the dry food that is fed is old or musty, an essential fatty acid deficiency will manifest as a dry dull haircoat.

4.5.4. Carbohydrate

Rabbits have little requirement for carbohydrate. However, many pet and house rabbits are fed starchy and sugary treats regularly. This causes obesity and encourages enterotoxaemia. The 'Starch Overload Theory' suggests that excess carbohydrates will exceed the stomach's capacity to absorb them, and these carbohydrates provide fuel in the caecum for the multiplication of E. coli and Clostridia.

4.5.5. Vitamins and minerals

During the process of digestion the microflora in the caecum synthesise vitamins B, C and K. These are then returned to the rabbit in the caecotrophes ingested during caecotrophy. Vitamins A, D and E are included in the pelleted ration. Concentrated ration should not be stored for longer than 3 months as the vitamin content will deteriorate over time. Vitamin B and K supplementation must be considered for rabbits that are unable to practise caecotrophy.

Vitamin A

This should be incorporated into the average concentrate ration at a rate of 10 000 iu/kg. VitaminA deficiency will result in infertility, abortions, resorption and increased neonatal mortality. Vitamin A deficiency or excess can cause hydrocephalus. Alfalfa is rich in vitamin A, and if fed alongside vitamin A rich pellets may also lead to the development of hydrocephalus. Hepatic coccidiosis will lower the liver levels of both vitamins A and E.

Vitamin C

This vitamin is also synthesised during the process of digestion. However vitamin C supplementation may be of benefit in the treatment of respiratory disease, and in the prevention of enterotoxaemia, as it may inhibit toxin production. It can be given at a dose of 50±100 mg/kg daily. Overdosage does not occur as any excess is excreted through the kidneys.

Vitamin D

This is incorporated into the average concentrate ration at a rate of 900 iu/ kg. Excess vitamin D may be a factor in excessive calcium uptake, and the development of dystrophic calcification of renal and aortic blood vessels (at levels of 2300 iu/kg of feed) although the rabbit has a unique calcium metabolism (see calcium).

Vitamin E

This is incorporated into the average concentrate ration at a rate of 50 mg/kg. Vitamin E deficiency can lead to reduced fertility, and additional vitamin E can be added to the diet by supplementation with wheatgerm. Vitamin E deficiency can cause muscular dystrophy and hind leg paralysis in young rabbits. These symptoms may be seen if less than 16 mg/kg vitamin E is incorporated into the diet. Blood biochemistry in a case of hypovitaminosis E will reveal a raised creatine phosphokinase (CPK).

Calcium

The rabbit has an unusual calcium metabolism. The level of plasma calcium is not regulated by vitamin D and parathyroid hormone as in other mammals, but varies directly with the level of dietary calcium. Excess calcium is excreted through the urinary system, making bladder and kidney stones a common occurrence if diets are high in calcium. Metastatic calcification can also occur with high calcium diets. Conversely, as much of the dietary calcium is held in the pelleted portion of the ration, rabbits that are selective feeders and leave their pellets may have a low calcium intake. This leads to osteoporosis, particularly evident in the vertebrae, and jaw bones. Poor mineralisation of the teeth, and acquired malocclusion are consequences of a diet too low in calcium. Osteoporosis of the vertebrae leads to vertebral fractures. The daily requirement for calcium of a medium-sized pet rabbit is approximately 510 mg.

Copper

Copper is added in the form of cupric sulphate to the concentrate ration at a rate of 4±30 mg/kg. It is thought to reduce enteritis and increase weight gain in young stock. There are no toxic effects from feeding this level of copper.

4.6. Feed Regimes during Different Life Stages

The nutritional requirements of rabbits, as in case of other mammals, vary according to age and productive performance. To obtain effective feeding efficiency, diets should be formulated to meet the needs of animals of a particular age or stage of production.

(a) Creep diets

Creep diets are those diets which are fed to babies since requirement for growth is highest in them. This is decreased with increase of age. The baby rabbit is capable of a much greater growth rate than is commonly observed with this diet. In view of the higher cost of creep diet, creep feeding does not appear to be an economically sound practice in rabbit production now a day.

(b) Weaning diets

Weaning is the period when animals diet changes from mild to solid feeds. It is suggested that at weaning, a high-fibre, low starch diet might be beneficial, followed by a switch to a high starch diet two weeks later; at which time the capacity of the animal to digest starch might be higher. For maximum production efficiency, a feeding system using highly palatable, high fibre diet at weaning with a switch to a high carbohydrate fattening ration may be beneficial.

Diet for Fryers (4-12 Weeks old)

Ingredients		Composition (%)
1.	Alfalfa meal	54.0
	Wheat mill run	36.5
	Soyabean meal	6.0
	Molasses	3.0
	Salt (Trace, mineralized)	0.5
		100
2.	Alfalfa meal	40.0
	Ground oats	27.5
	Ground barley	20.0
	Soyabean meal	9.0
	Molasses	3.0
	Salt (Trace, mineralized)	0.5
		100
3.	Alfalfa meal	40.0
	Ground barley	25.0
	Wheat mill run	23.5
	Cottonseed meal	8.0
	Molasses	3.0
	Salt (Trace, mineralized)	0.5
		100

(C) Gestation/ Lactation

Lactating does have higher requirements for proteins, energy, calcium and phosphorus than do fryers. For maximum production atleast 18% CP Is required. Since, does in commercial leads are simultaneously pregnant and lactating, the same diets can be used for both lactation and gestation. During periods when does are not with litters, restricted feeding should be practices to avoid obesity.

Peak lactation in rabbits occur 21 days post kindling. In does bred 24-48 hr. postpartum, milk production declines rapidly after 21 days of lactation, and the mammary glands prepare for the initiation of a new lactation period.

(D) Pregnant females on daily basis

Fodder	Does	
	(Second half of pregnancy)	
	In winter (g)	In summer (g)
Mixture of concentrate	80	70
Hay	200	50
Steamed potatoes	100	60
Carrots, beet roots	300	-
Green forage	-	900
Mineral mixture	3-5	2-3
Sprouted grain	3	-
Cod liver oil	1	1

(E) Feeding schedule for suckling females (ration in kg/head/day)

Female	Fodder	Ration in winter	Ration in summer	Remarks
Doe	**Concentrate Potatoes**	0.9 0.1	0.60.08	Supplement of 8g concentrate, 10g potatoes, 100g succulent fodders (in summer)
	HayRoot crops Green	0.250.4	0.05	10g concentrate, 15 g potatoes, 20g succulent fodder, 10 g hay (in winter)

(F) Ration for lactating doe

Ingredient	Amount (%)
Alfalfa	40.0
Ground barley	20.25
Wheat mill run	20.0
Soyabean meal	14.0
Molasses	3.0
Fat	1.5
Dicalcium phosphate	0.75
Salt	**0.5**

(G) Complementary diets

In certain conditions, particularly with the small scale rabbit production, it may be desirable to feed hay or greens free of choice and supplement this diet with a restricted quantity of high energy, high protein concentrate.

4.7. Feeding Schedule

A pregnant doe may consume 200 to 250g of balanced pelleted feed combined with good green grasses. Grass can be provided @100g/day. A balanced nutritive feed may have the following percentage of ingredients.

Nutritional requirement of a pregnant doe

Crude protein	17 %
Crude fat	2 %
Crude fibre	14 - 17 %
Methionine	1.0 %
Arginine	0.3 %
Cystine	0.3 %
Lysine	0.5 %
Vitamins (ADEK)	Adequate
Minerals	1.5 %
D.E.	2600 kcal/kg
T.D.N.	600

4.8. Feed Consumption

Daily feed consumption is about 5% body weight. Daily water consumption is about 10% body weight. Lactation does will require more water and food. A rabbit keeper should make the diet schedule accordingly. Clean fresh water should be made available at all times.

Approximate Feed and Water Consumption for Broiler Rabbit

Age (Weeks)	Weight (g)	Water consumption (g)	Feed consumption (g)
6th	9200	133	84.5
7th	1200	177	113
8th	1510	220	140
9th	1810	243	153
10th	2070	258	161
11th	2300	267	165
12th	2510	276	168
Fully matured	4000	300	150-170

4.9. Nutrient Requirement for Rabbits

Total %	Protein Digestible	Protein %	Fat %	Fibre %	Carbohydrate (Digestible %)	Total digestible nutrient %
Maintenance	12	9	1.5 - 2.0	14 - 20	40 - 45	50 - 60
Growth	16	12	2.0 - 4.0	14 - 16	45 - 50	60 - 70
Pregnancy	15	11	2.0 - 3.0	14 - 16	45 - 50	55 - 65
Lactation	17	13	2.5 - 3.5	12 - 14	45 - 50	65 - 75

4.10. Feeding of Rabbit

Milk is the only food for baby rabbit upto the age of 15-21 days. Baby rabbits that are able to suckle survive, others may die. The rabbit keeper should ensure milk production in does by providing plenty of food and water. The young rabbit will start to eat solid foods (concentrates, grass) along with mother's milk from 15-21 days. After 21 days of so, they eat more solids and suckle less. From weaning time onwards the rabbit should be provided with more green feed, vegetables and concentrates.

The ration of a rabbit can be prepared at home from locally available ingredients.

4.11. Homemade Feed for Non Breeding Rabbit

Ingredients	Weight (G)	Percent (%)
Diet A		
Alfalfa hay	127	64
Oats	29	15
Barley	29	15
Soya meal	15	6
Total	**200**	**100**
Diet B		
Hay (Mixed quality)	8040	1613
Hay (Good quality)	100	31
Greens kitchen scrap	100	31
Oats	30	9
Total	**350**	**100**

4.12. Homemade Diet for Pregnant Rabbit

Ingredient	Weight	Percent
Alfalfa hay	114	64
Barley	24	14
Sorghum grain	24	14
Soyabean meal	15	8
Total	**177**	**100**

4.13. Homemade Diet for Lactating does

Ingredient	Weight	Percent
Alfalfa hay	109	21
Wheat	156	31
Sorghum grain	56	11
Soyabean meal	188	37
Total	**509**	**100**

4.14. Time of Feeding

It will be a wised proposition if the food is offered in a particular time of the day. All concentrate ration may be provided between 7 am and 5 pm. Fresh and bulky feed may be offered in the evening since the rabbit is more active during that time. All kinds of food should be fresh and free from dirt.

4.15. Composition of Standard Pellet Mixture

Ingedients	Growing Stage	Pregnant Stage	Lactating Stage
Maize	35	15	50
Barley	-	10	-
G.N.C	17	11	20
Fish meal (Dried)	5	5	5
Wheat grain	33	50	20
Molasses	10	9	5
Total	**100**	**100**	**100**

4.16. Some Consideration on Feeding

1. Fresh grasses or legumes may be included in the ration to a level of 70%.
2. 50% legumes plus 50% grasses may afford minimum requirement for resting does and bucks.
3. Dustiness of mash may be avoided by adding water to moisten the feed slightly.
4. Ensurement may be made to avoid acidity (sour) since sour feed is not liked by rabbits.
5. Clean and fresh water may be made available in all seasons.
6. Does should not be fed with heavy diet right after kindling.
7. Feed of does may be increased after 5-7 days of kindling.
8. Abrupt change in quality and quantity of feed should be avoided.
9. Molasses @ 5%may be added to the diet to increase acceptability of the diet.

10. A small quantity of hay/ straw may be fed in addition to the compound feed to prevent boredom and to provide bulk.
11. Rapeseed oil meal may be heated before inclusion in the diet. It may be fed below 15% levels in diet.
12. High calcium diet may interfere with the acceptance of off springs by the does and thus may affect the growth of the litters.
13. Dry does, bucks and replacement stock may be fed once a day at the rate of 100-200g of pellet feeds.
14. Young rabbits in growing stages and lactating does may be fed free of choice by keeping grains in the feeder at all times.
15. Good quality legumes may be kept before the rabbits in addition to pellet ration.
16. Carrots, green grass, spinach, turnip, Lucerne, berseem etc may be provided as good succulent feed for rabbits.
17. Other feed stuffs such as kitchen wastes, spoilt milk, damaged fruits may be fed by small farmers but they should be fed with caution considering their deleterious effects.

4.17. Utensil for Feeding

A good number of watering utensils and feeders are used in the rabbit cages. The feeder should be kept at 5 - 8 cm height so as to minimize contamination of feed by faeces, urine or water. Open food and water bottles are less hygienic than hoppers and water bottles. Earthen and aluminum bowls could be used since they are cheaper from economic stand point. Hoppers should be made as such so that the food cannot spill out. The water bottles if used should be cleaned properly. Filling and cleaning of water bottles in large farms may require considerable time; therefore, automatic watering systems are gaining importance in big rabbitry. The feeder, whatever may be the nature, should be cleaned properly each day before use.

❑❑❑

Chapter – 5

Reproduction in Rabbits

5.1. Anatomy of the Genitals

5.1.1. Male reproductive system

Testes in the male are oval-shaped present within the scrotum which remain in communication with the abdominal cavity, where they were at birth. The rabbit is actually able to withdraw its testes when frightened or fighting with other males (Fig. 5.1). The testicles descend at about two months. The short, back-slanting penis points forward when erect.

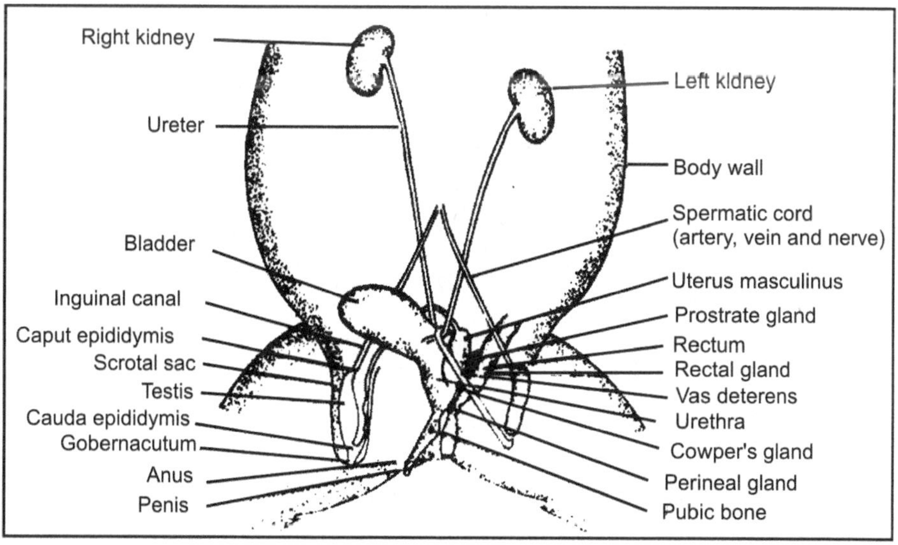

Fig. 5.1: Male reproductive system of rabbit

Glands associated with reproduction are:

1. Prostrate gland
2. Cowpers gland / Bulbourethral gland
3. Perineal glands
4. Rectal glands

5.1.2. Female reproductive system

In the female, ovaries are oval-shaped and do not exceed 1 to 1.5 cm. Beneath the ovaries is the oviduct, made up of the duct, the ampulla and the isthmus. Although outwardly the uterine horns are joined at the back into a single organ, there are actually two independent uteri of about 7 cm, opening separately through two cervical ducts into the 6 to 10 cm vagina. The urethra opens midway along the vagina a.t the vaginal vestibule (Fig. 5.2). The glands of Bartholin and the preputial glands can be identified. The whole is supported by the broad ligament attached at four main points under the vertebral column.

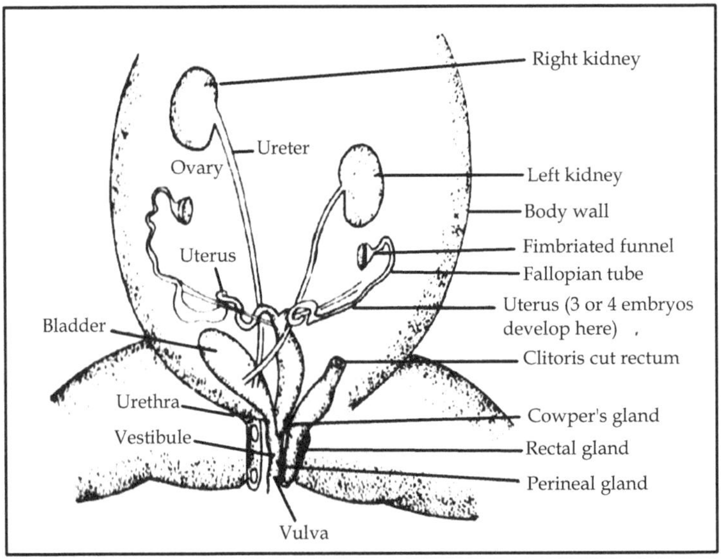

Fig. 5.2: Female reproductive system of rabbit

5.2. Male Reproduction Physiology

5.2.1. Male gonad development and puberty

The gonads begin to differentiate on the 16th day after fertilization. After birth the testes develop less quickly than the rest of the body.

From the age of five weeks they begin to grow very rapidly. Accessory glands undergo a similar development, but at a more even rate and are less precocious. Spermatogenesis begins between days 40 and 50. The testicular tubes become active at about 84 days. The first spermatozoa are present in the ejaculate at about 110 days. Sexual maturity, defined as the moment when daily sperm production ceases to increase, is reached at 32 weeks by New Zealand White rabbits in temperate climates. However, a young buck in these same conditions can be used for reproduction from the age of 20 weeks. Indeed the first manifestations of sexual behavior appear at days 60 to 70 when the rabbit makes it first attempts at riding. Coitus may occur for the first time at about 100 days, but the viability of the sperm cells is very weak or nil in the first ejaculates. So first mating should be timed for age 135 to 140 days. The onset of puberty varies from breed to breed, but conditions in the rabbitry also play an essential role, particularly feeding, which is even more important than climate.

5.2.2. Sperm production

The volume of semen ejaculated is about 0.3 to 0.6 ml. Concentration is evaluated at 150 to 500 x 106 spermatozoa per ml, but both volume and concentration are liable to vary. False mountings, one or two minutes before copulation, increase the concentration of the ejaculate. In two successive servicings the first acts as a preparation for the second, which is less voluminous but more concentrated. During subsequent matings the volume of the ejaculate decreases, while concentration increases between the first and the second ejaculate and then diminishes. The total number of spermatozoa per ejaculate follows the same trend. Maximum spermatozoa production is obtained by using the buck regularly once a day. If the buck is used regularly twice a day, each ejaculate has only one half the concentration of spermatozoa. On the other hand, if bucks service several times a day, one day a week, the three or four ejaculates may be concentrated enough to effect fertilization. Further ejaculates contain very few spermatozoa and cannot effect fertilization often enough to be worth while. Daily spermatozoa production is roughly 150 to 300 million, independent of the rate of ejaculation. The maximum epididymis reserve is only one to two billion spermato-zoa, only partially mobilizable for repeated ejaculations.

5.3. Female Reproduction Physiology

5.3.1. Female gonad development, puberty and sexual maturity

As in the male foetus, sexual differentiation takes place on the 16th day after fertilization. Ovogonial division begins on the 21st day of foetal life and continues until birth. The first follicles appear on the 13th day after birth, and the first antrum follicles at about 65 to 70 days. Does are able to mate first at 10 to 12 weeks, but as a rule this will not produce ovulation. The onset of puberty varies greatly with:

- The breed: sexual precocity is more developed in small or medium breeds (four to six months) than in large breeds (five to eight months).
- Body development: precocity goes hand in hand with rapid growth. Does fed ad lib reach puberty three weeks earlier than other does of the same strain receiving only 75 percent of the same daily feed. The body development of the latter is also delayed by three weeks.

Does generally reach puberty when they have grown to 70 to 75 percent of their mature weight. However, it is usually preferable to wait until they reach 80 percent of their mature weight before breeding them. These relative weights should not be considered absolute thresholds for all rabbits, but rather limits applicable to the population as a whole. Sexual behaviour (acceptance of mating) appears long before the ability to ovulate and bear a litter. Such behaviour should not be regarded by the breeder as a sign of puberty, but as prepuberty play.

5.3.2. Oestrus cycle

In most domestic mammals ovulation takes place at regular intervals when the female is in heat or oestrus. The interval between two periods of oestrus represents the length of the oestrus cycle (four days for rats, 17 for ewes, 21 for sows and cows). The female rabbit, however, does not have an oestrus cycle with regular periods of heat during which ovulation will occur spontaneously. Does are considered to be in oestrus more or less permanently. Ovulation occurs only after mating. A female rabbit is therefore considered to be in heat when she accepts service and in dioestrus when she refuses.

A red vulva is a strong indication, though not a proof, of oestrus. A doe in heat assumes a characteristic pose, called **lordosis**, with the back arched downwards and hindquarters raised. A doe in dioestrus tends to crouch in a corner of the cage or exhibit aggression towards the buck. The sexual behaviour of a female rabbit is thus very special. She has no cycle and can stay in heat for several days running.

On the ovary, follicles not having evolved to the ovulation stage through lack of stimulation undergo regression and are replaced by new follicles, which remain for a few days in the pre-ovulating state and may then in turn regress. In most mammals the progesterone secreted during gestation inhibits oestrus and the pregnant female refuses to mate, but a pregnant doe may accept mating throughout the gestation period. Indeed, in the second half of pregnancy this is the most common behaviour. A breeder cannot therefore use the sexual behaviour of does as an indication of pregnancy. Mating occurring during gestation has no dire consequences for the embryos. Unlike the phenomenon observed in the female hare, superfoetation (two simultaneous pregnancies at two different stages of development) never occurs in rabbits.

5.3.3. Ovulation

Ovulation is normally induced by the stimuli associated with coitus and occurs ten to 12 hours after mating. Ovulation can be induced artificially by various techniques which are:

1. Mechanical stimulation of the vagina.
2. Injections of luteinizing hormones (LH) or LH releasing hormones (LHRH).

5.4. Fertilization and Gestation

At the moment the ovary follicles are ruptured the oviduct pavilion or infundibulum covers the ovary. When liberated the ovocytes are sucked in by the pavilion. The ovocytes are in fact fertilizable from the moment they are liberated, but they are not actually fertilized until about an hour and a half after release. The sperm is deposited by the male in the upper part of the vagina. The spermatozoa make their way upwards rapidly. They can reach the fertilization area (in the distal ampulla, near the isthmus) 30 minutes after coitus. During their journey the spermatozoa undergo a maturing process which enables them to fertilize the ovocytes. Of the 150 to 200 million spermatozoa ejaculated, only two million (1 percent) will reach the uterus. The rest are defeated by obstacles at the cervix and uterotubal junction. The egg reaches the uterus 72 hours after simultaneously lactating and post-partum pregnant doe (fertile mating 24 hours after giving birth), late embryonic mortality is increased with respect to that observed in a simply pregnant doe under the same circumstances.

5.5. Pseudo Pregnancy

Liberated ova which are not fertilized may occasion a pseudopregnancy lasting 15 to 18 days. At first the corpus luteum and uterus develop as in an ordinary pregnancy, but they do not reach the size or the level of progesterone production of the corpus luteum in pregnancy. Towards the 12th day they regress and disappear under the action of a luteolytic factor secreted by the uterus, undoubtedly prostaglandin. The end of pseudopregnancy is marked by the maternal behavior of the doe and nest-making, linked to the swift drop in blood progesterone.

5.6. Kindling

The act of parturition in rabbits is known as kindling. At the end of gestation the doe makes a nest for the litter with her own fur and materials she has available such as straw and shavings. This behaviour is linked with an increase in the oestrogen / progesterone ratio and with the secretion of prolactin. The doe does not always make a nest, or she may kindle outside the nesting box. Kindling lasts from 15 to 30 minutes, according to the size of the litter. Litter size varies as much as from one to 20 young. Kindling involves following steps:

1. Contractions of the muscular walls of the uterus.
2. Widening of its opening.
3. The busting of amnion and escape of amniotic fluid.
4. Expulsion of fetus.

After parturition the uterus retracts very quickly, losing more than half its weight in less than 48 hours.

5.7. Artificial Insemination in Rabbits

5.7.1. Semen collection

A doe in heat is put into the buck's cage. The operator holds the artificial vagina with its collection tube between the rabbit's paws. The artificial vagina is kept at a temperature of about 40° to 42°C prior to use, so that it will be at 39°C, the normal vaginal temperature of a doe, at the moment of use. Ejaculation usually takes place immediately following the presentation of the doe. The semen is then diluted five to ten times, perhaps in physiological salt solution, within 30 minutes after semen collection, or with a special diluent if it is to be applied within 12 hours. It is possible to freeze the semen, but the poor performance of frozen semen relegates this technique.

5.7.2. Insemination

The semen can be packaged in 0.5 ml pellets or presented in 20,50 or 100 0.5 ml flacons for insertion with a glass cannula. Two techniques co-exist:

1. An insemination gun covered with a single-use sheath
2. The glass (or throwaway plastic) cannula.

Both techniques have their partisans and their detractors and for both the diluted semen must be delicately inserted deep into the rabbit vagina.

As ovulation is not spontaneous in rabbits, intramuscular injection of an artificial analogue of GnRH (gonadoreline 20 busereline 0.8n) is used to provoke ovulation at the moment of insemination. AI in rabbits involves a dual intervention: insemination and the injection of an ovulation- producing hormone.

5.8. Lactation

Milk synthesis depends on prolactin, a lactogenic hormone. During pregnancy prolactin is inhibited by the oestrogens and by progesterone. At parturition there is a rapid drop in the progesterone level. As oxytocin is freed the action of the prolactin is stimulated and permits the milk to mount in a predeveloped gland. Milk is let down as follows:

1. The doe comes into the nest box to nurse her litter.
2. The stimulus of nursing provokes the secretion of oxytocin, inframammary pressure mounts, the milk is let down and the young suckle.

The amount of oxytocin secreted is proportional to the number of young feeding. But the doe sets the number of feeds: just once in 24 hours. Suckling alone will not provoke the secretion of oxytocin; the mother must want to nurse.

5.9. Aspects of Milk Production

Doe's milk is much more concentrated than cow's milk except for the lactose component. After the third week of lactation the milk becomes markedly richer in proteins and especially fats (up to 20 to 22 percent). The already low lactose content tapers off to almost zero after the 30th day of lactation. Daily milk production increases from 30 to 50g in the first two days to 200 to 250 g towards the end of the third week of lactation. It then drops rapidly. The decrease is even swifter if the doe has been fertilized immediately after kindling. The lactation curve varies from doe to doe, especially with regard to duration.

Measuring the young rabbits' weight at 21 days gives a fairly good estimate of total lactation, as milk production between days zero and 21 is closely correlated with total milk production (r + 0.92). An important point is that the doe's milk output increases with litter size but the baby rabbits get less milk each than they would in a smaller litter. Depending on genetic type, milk production will not increase above eight to 12 baby rabbits.

5.10. Reproduction and Environment

5.10.1. Lighting

In males exposed to artificial lighting for only eight out of 24 hours significantly more spermatozoa are present in the gonads than in those exposed to light for 16 hours. Does, however, are far more opposed to mating with only eight hours of light than they are with 16. For both males and females 12 hours of light a day produce average results.

5.10.2. Temperature

High temperature (>30 - 33°C) has following effects on the reproduction of rabbits:

1. Decrease in the volume and concentration of ejaculates in males.
2. Sperm motility in males is affected.
3. Bucks' sexual urge is reduced.
4. Lower prolificacy in does.
5. Increased embryonic mortality.

5.10.3. Season

Season is usually analysed in terms of the combined effects of light and temperature. In tropical climates the temperature effect seems to be dominant but an effect due to variations in the length of day-light cannot be excluded. The reproduction cycles of the European wild rabbit are strongly influenced by the season. Does breed from the end of winter until early summer. The reproduction period can be longer or shorter, at either end, according to both temperature and availability of feed.

5.11. Rates of Reproduction

The physiological features of the male and especially the female are such that the breeder has great latitude in choosing a reproduction method. The goal is to increase doe productivity and reduce inputs. Productivity is defined as the number of young per doe per unit of time and it depends on:

- The interval between successive kindlings
- Litter size at birth
- The survival rate of the young.

The crucial factor in increasing productivity is shortening the kindling-to-mating interval. This means non-productive periods must be reduced to the minimum.

There are three basic rates of reproduction (Fig. 5.3) are:

1. Extensive reproduction rate
2. Semi-intensive reproduction rate
3. Intensive reproduction rate

1. Extensive reproduction rate

The breeder fully utilizes the does' maternal instincts by allowing them to nurse their young for five to six weeks, rebreeding them soon after weaning. Does are therefore serviced once every two and a half months. Later weaning is in no way advantageous except for fryer production - very young animals which can be sold at eight weeks and have not undergone weaning shock. The mother can be serviced before weaning, about five or six weeks after kindling, which allows two and a half months between litters. Where the quality or quantity of the feed is not up to standard, it is preferable to wean rabbits at about 40 days. At the same time the breeder should slightly lengthen the resting period between weaning and rebreeding so the doe can build up her reserves again. In any case, weaning later than six weeks offers no particular nutritional advantage. The milk produced by the doe after this period provides at most 3 to 5 percent of the young rabbits' daily feed intake.

2. Semi-intensive reproduction rate

The breeder has does serviced 10 to 20 days after kindling and the young are weaned at four to five weeks. There is no real contrast between pregnancy and lactation for does. For 10 to 20 days the doe is newly pregnant while still nursing. The most important phase of embryo development takes place during the slump in milk production (milk production may even have ceased), so there is no real competition between the demands of gestation and lactation. As these does never have a resting period they need sufficient and well-balanced concentrate feed.

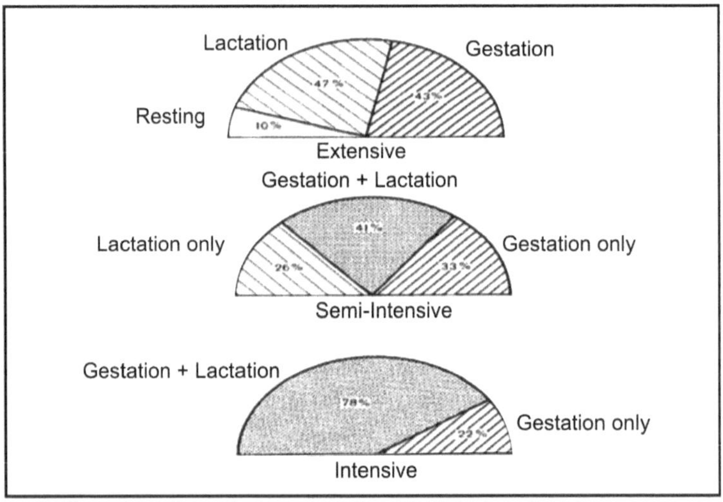

Fig. 5.3: Different rates of reproduction in rabbit

3. Intensive reproduction rate

The breeder has the does reserviced just after kindling, taking advantage of the fact that they are then on heat. Weaning should take place at four weeks at the latest, usually at 26 to 28 days.

There are three main techniques:

a) Servicing the same day or the day after kindling: the true postpartum rate.

b) Servicing scheduled for a specific day, generally three or four days after kindling

c) Ad lib mating: A buck left together with postpartum does will serve them several times during the 48 hours following kindling. This is the natural rhythm of wild rabbits. To arrange ad lib mating, breeders have worked out two types of rabbit housing which are:

i. Corridor-collar type

The does live in individual cages. They have a broad collar around their necks to prevent them from leaving the cage through the calibrated opening leading into a communicating corridor. The buck, however, has free access (at least temporarily) to the does' cages and can mate whenever the doe is ready.

ii. Group system

A buck and perhaps ten does live together in the same cage. They can mate at the optimum times. However, special arrangements must be made to curb the natural tendency of females to kill the offspring of other does when they themselves are lactating or ready to kindle.

❑❑❑

Chapter – 6

Management of Rabbitry

6.1. The Production Cycle

As ovulation in does is induced by mating and the females are generally kept in different cages from the males, it is the breeder who determines the reproduction rate of the unit. These rates vary from one or two litters a year under the most extensive management to eight to ten litters in an intensive management system. Does are remated either immediately after kindling (intensive system) or about ten days later (semi-intensive). In the semi-intensive system, the does are first presented to the buck at four and a half months. They are then mated 10 to 12 days after the birth of each litter. Weaning takes place at 30 to 35, or even 37 to 38 days. Extensive systems are characterized by a long delay between kindling and mating, and perhaps even until weaning. For example, the young may be weaned at 56 days and the doe mated after weaning.

At weaning the young are separated from the doe. The duration of fattening varies, depending on the carcass weight required and the growth rate possible in the feeding and production conditions of the rabbitry. When weaning takes place at one month, the fattening period is seven weeks. The rabbits weigh 2.3 to 2.4 kg (live weight) when they are ready for the market. When weaning takes place at two months a four month fattening period is required.

6.2. Mating

Servicing is always done in the buck's cage. The breeder checks the doe's health at this time to make sure she has no respiratory disorder, sore hocks, etc., or that she is not too thin. A red vulva is a promising but not infallible sign (80 to 90 percent chance of mating success). A buck can fertilize a doe with a white vulva, but the success rate is only 10 to 20 percent. When the doe has accepted mounting and the buck has serviced her the breeder removes the doe and puts her back in her own cage. Altogether this should not take more than five minutes. If the doe refuses to mate, the breeder can try to present her to another buck, as a last resort leaving her for 24 hours in the buck's cage but then cannot be sure that mating has taken place. It is better to mate the animals in the morning or evening, to avoid the hottest hours of the day.

Double mating is another method of mating in which doe is mated twice in succession at intervals of 10 to 15 minutes, either by the same male or by two different males.

A similar technique is to leave the female in the male's cage for 15 to 20 minutes after mating has first taken place. These techniques allow a slight increase in the percentage of pregnant does (4 to 6 percent, roughly). The drawback, however, is that it considerably increases the number of matings per male and in this system no male can be mated to more than two females each week without jeopardizing the outcome through over exploitation of the buck.

In intensive breeding one buck can serve seven or eight does. In the extensive system one buck can serve 10 to 15 does. The buck, however, should not be used more than three or four days a week, and not more than two or three times a day, which means no more than six ejaculations per week. When the size of the unit permits (at least 50 does), one or two reserve bucks are kept.

If a balanced pelleted feed is used the bucks should be fed from 120 to 180 g per day, depending on their weight. Bucks are first mated at about five months. If production conditions are not optimum the first mating will be delayed until the animals reach 80 percent of their adult weight.

6.3. Determining Pregnancy

An effective way of determining pregnancy is to detect the embryos in the doe's uterus by palpating the abdomen. This operation should be carried out between 10 and 14 days after mating. It is not effective if performed earlier (before the ninth day), while after the 14th day the operation is more delicate and there is a risk of provoking abortion. The breeder must palpate the doe gently and expertly in order not to cause an abortion. If palpation shows the doe to be empty she is presented to the buck again as soon as possible.

Presentation of the doe to the buck as a test of pregnancy is pointless, though not dangerous. Indeed, a large proportion of pregnant does accept mating and some empty does refuse. Nor is doe live weight an indication of pregnancy, because weight fluctuations depend on too many factors.

Preparations for kindling (supervision, nest box, changing bedding material, etc.) should be made for all the mated does from the 27th to the 28th day after mating if they have not been palpated, but where palpation has been practised regularly the preparations are restricted to does found to be pregnant. A pregnant doe that is not nursing a litter will be rationed if the breeder uses pelleted feed. The daily ration for medium-size does will be about 150 g (35 to 40 g / kg live weight). If the doe is nursing a litter at the same time she will be fed ad lib.

6.4. Kindling

Kindling should take place in quiet, hygienic surroundings. The breeder's presence is not required, but the nests should be checked as soon as possible after kindling. This operation is easy and there is no risk to the young. It can be performed right after kindling, provided the mother is removed. Any dead animals and any foetal sacs the doe has not eaten should be removed.

A nursing doe needs considerable nutrition and from the time of parturition she should be fed ad lib. Drinking-water is very important in the days leading up to and following parturition.The doe will nurse her young once a day, usually in the early morning. Strict preventive hygiene is more important than ever at this period.

6.5. Fostering

Fostering by other doe may be required when the rabbit is unable to provide milk to her own litters. The only feed upto the age of 20 days in milk. Therefore, new born rabbit at birth can be transferred to foster mother even within 1 to 2 days of birth following the death of own mother. Few rules to be followed are:

- No more than three or four young rabbits should be given to a foster mother.
- The maximum age difference between the foster doe's litter and the fostered young should be 48 hours.
- Fostering should take place within three days of kindling.

6.6. Weaning

Weaning is the time when the young one is separated from the doe. Weaning can take place when the rabbit's live weight tops 500 g or after 4 weeks of age. The young rabbits begin to eat solid feed at 18 to 20 days and at 30 days the doe's milk provides no more than 20 percent of the daily dry-matter intake. Depending on the rate of reproduction chosen, weaning should take place no later than two or three days before the doe's next kindling.

There are two methods adopted for weaning of rabbits which are:

1. All rabbits in the litter are withdrawn at the same time and placed six to eight per cage in the area set aside for fattening.
2. Alternatively, the doe may be removed from the cage and the young rabbits left. This method reduces postweaning stress for the young rabbits.

Management must be geared to group rearing. If the young rabbits are moved, the cages must be very clean and the litters should be kept together, if possible, for uniformity. The alternative is same-age cages (maximum age difference one week) with all rabbits put in the cage the same day. Rabbits soon establish a social hierarchy in the cage and any new introduction is a source of conflict. During the transfer operation the breeder checks the health of the young rabbits, culling any that are undersized or sick.

6.7. Stock Renewal

Renewal mainly takes the following two forms:

1. The introduction of male or female breeding animals for direct replacement of does or bucks which have been culled or died. These are called "Parental Renewal Stock" (Direct parents of rabbits intended for sale).
2. The introduction of grandparents. Here, the parental rabbits are born within the breeding establishment, the progeny of bucks and does of special complementary lines which live and produce in the same establishment and have a reserved place in it. These grandparent rabbits are in turn replaced by rabbits direct from the selection centre, but in much smaller numbers than needed for the direct renewal of parental animals.

Whatever the genetic type of rabbits brought into the establishment for the renewal of breeding stock, they should be brought in at a fairly early age. The young rabbits adapt much better than those introduced at the age of 8 to 11 weeks, and particularly four months or older.

6.8. Castration of Young Male Rabbit

Castration in rabbits is a simple operation, though it usually requires two people. The rabbit is held on its back on the lap of an assistant who holds both legs on the same side in each hand. The operator presses the rabbit's belly from front to back to make the testicle appear. Using a razor, he makes a deep incision, cutting the skin of the pouch and also the testicle (Fig. 6.1). At this moment, by a reflex movement, the testicle comes out It now remains only to cut the "threads" fixing it to the abdomen. Next disinfect with iodine. There is no need to stitch the wound as it heals in five or six days. Bleeding to any extent is very rare

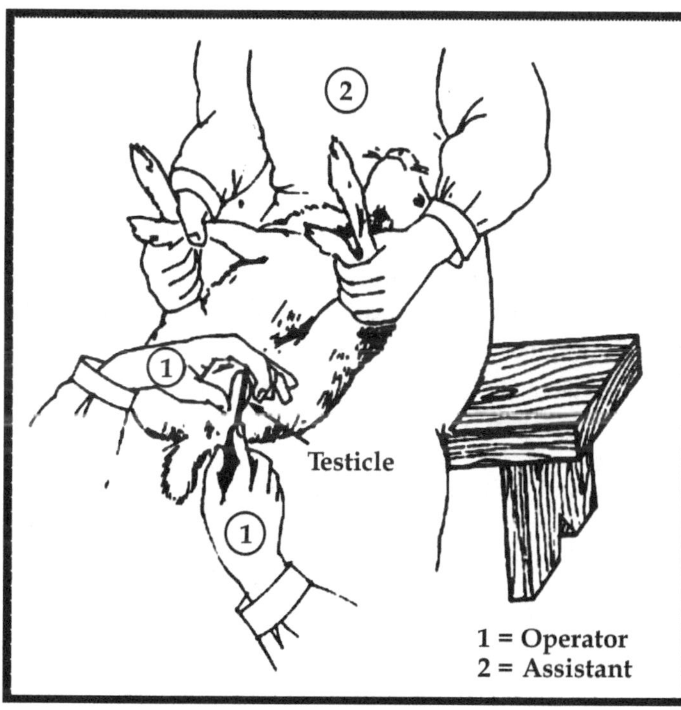

Fig. 6.1: Method of castration of young rabbits

However, for a rabbit of four months and over it is best to crush the threads rather than cut them. There are castrating pincers which crush the threads, eliminating the need for an incision.

6.9. Handling Rabbits

Rabbits should be handled gently. They should be lifted by their ears as little as possible. Several techniques can be used to pick them up and hold them. A rabbit can always be picked up by the skin of the back. (Fig. 6.2)

Fig. 6.2: Correct way to pick up a rabbit

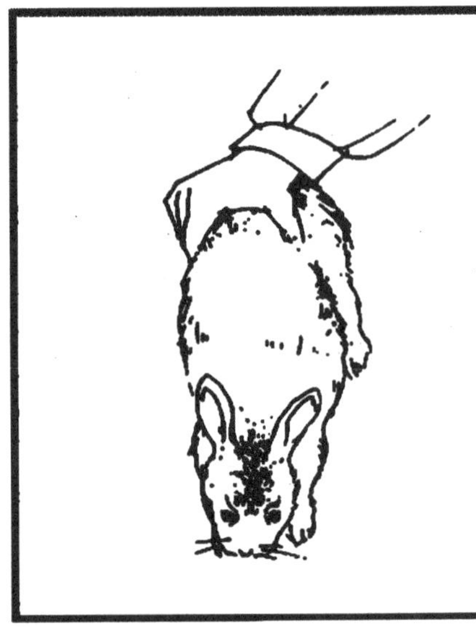

Fig. 6.3: Holding a young rabbit

For animals weighing under one kilogram, one method is to pick them up and carry them by the saddle just above the hindquarters, using thumb and index finger. (Figure 6.3)

If the animals are heavier it is best to take them by the skin of the back, but if they have to be transported or shifted for more than five or ten seconds they must either be supported with the other hand (Fig. 6.4) or be carried on the forearm with the head in the bend of the elbow. (Fig. 6.5).

If an animal struggles and the producer feels he cannot control it, it is best just to drop it so it will fall on all fours and then pick it up again correctly within two or three seconds. If handler keeps his hold on a struggling rabbit he risks some nasty scratches and can even break the rabbit's backbone.

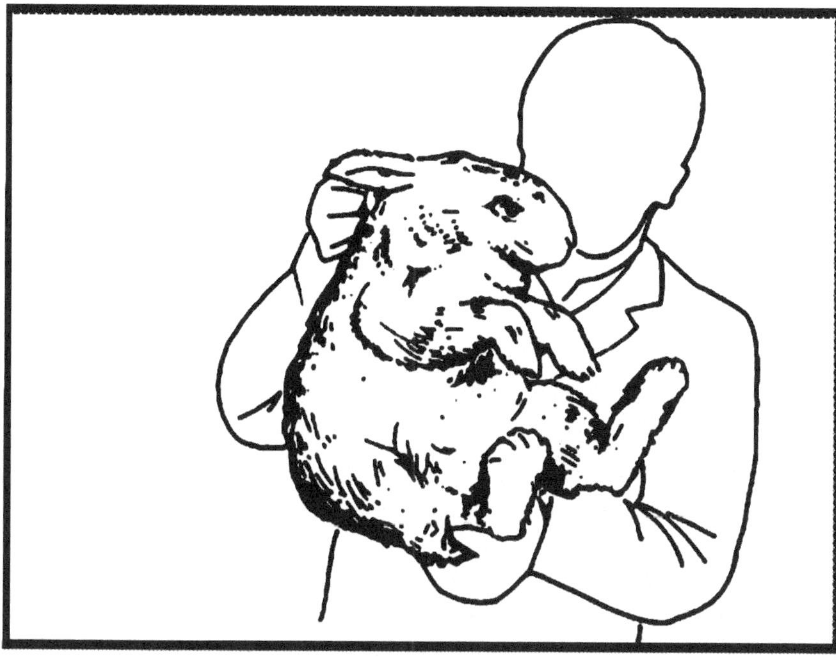

Fig. 6.4: Method of carrying a large sized rabbit

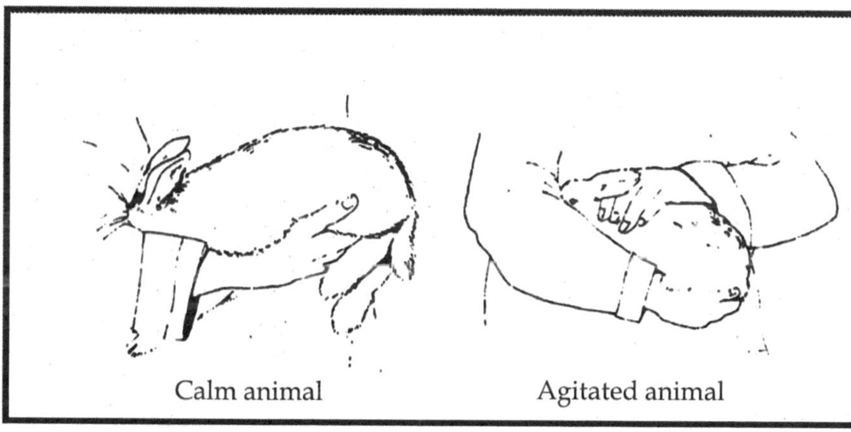

Fig. 6.5: Technique for carrying a rabbit on the forearm

6.10. Identification

Identification can be of two types:

1. Individual identification
2. Cage identification

6.10.1. Individual identification

In this method each animal is assigned a number. This number will appear on all documents concerning the rabbit and on the rabbit itself. There are three main ways of identifying rabbits individually which are:

i. **Method of using rings**: A numbered ring is attached to the hind leg just above the hock but the risk of losing the ring is high.

ii. **Method of using clips**: Numbered clips are attached to the rabbit's ears. These clips are made of metal or plastic and risk of loss is again high (Fig. 6.6).

iii. **Tattooing method**: Small holes are punched in the rabbit's ear spelling out numbers or letters and these are filled by rubbing in a special ink. A well-made tattoo lasts throughout the rabbit's lifetime. Although this method takes longer, it is the only one that is really sure. It can be done at weaning using special rabbit pincers, or on adult rabbits with sheep pincers.

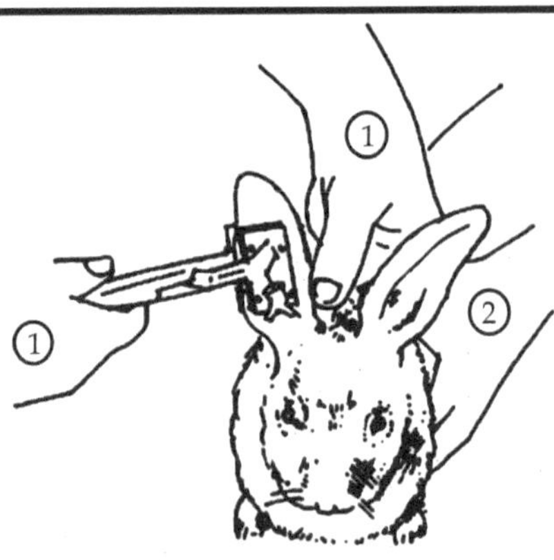

Note : 1 = operator's hand, 2 = assistant's hand.

Fig. 6.6: Technique of using clippers with movable numbers to tattoo on rabbit ear

6.10.2. Cage identification

In this method all the cages in the nursery section are numbered and this is the figure that will appear on the records. This method is much easier than individual identification so it is used in rabbitries which keep records but do not breed for selection purposes.

6.11. Record Keeping

6.11.1. In the nursery

A daily record book is maintained and followings operations are recorded in the nusery:

1. Mating days for each doe.
2. Outcome of palpation.
3. Size of each litter at birth.
4. Size of each litter at weaning.
5. Litter weight at weaning.
6. Amount of feed fed to compute the average feeding cost per weaned rabbit.

Doe card can be hooked to each cage, for calculating individual doe productivity. Similarly **buck card** can also used for recording individual buck performance.

6.11.2. Fattening records

A daily record book is maintained and followings operations are recorded:

1. First and last fattening days (sale or slaughter) of the animals in each cage.
2. Any mortality and the apparent causes.
3. Live weight when sold.
4. Number of animals marketed weekly.
5. Amount of feed eaten by fattening rabbits.

For selection purposes a litter card is used listing the following:

1. Weaning weight
2. Weaning date
3. Weight at slaughter
4. Date at sale or slaughter
5. Individual identification number of each rabbit.

❑❑❑

Chapter – 7

Diseases of Rabbit

Health care and preventive hygiene in rabbits are very important for the productivity of the rabbits and economical viability of a farm.

It is always better to prevent the disease rather than to get an animal treated, which is usually quite expensive. As such it is desirable that the rabbits are checked daily. The health and ailment signs can be recognized from the following things:

1. **Coat:** The coat of the animal should be smooth, glossy and unruffled.
2. **Eyes:** It should be bright and no discharge coming out.
3. **Movements:** The movement should be free and easy and the rabbit should look alert. It should not sit huddled/ crouched in the cage or near the watering bowl.
4. **Body Conformation:** It should be well fleshed at appropriate places, should neither be pot-bellied, skinny nor boney as felt from the backbone. They should not loose body weight.
5. **Temperature/Pulse/Respiration:** The pulse rate of healthy rabbit is about 130-150 per minute and normal body temperature is 37°C and respiration rate is 40-65 per minute.
6. **Feeding:** A normal rabbit should have good appetite and feed well. If it goes off feed it is a sign of illness.
7. **Droppings:** These should be normal and semi-hard in consistency. Loose droppings or with mucoid coating or diarrhea are signs of illness.

8. **Discharges:** There should not be any discharge from any natural orifices i.e., nostrils, mouth, eyes, ears, anus and teats.
9. **Sores and swellings:** The body surface of rabbit should not have any sore or show swellings of any area.
10. **Temperament:** The animal should be calm and composed and should not show any signs of restlessness, excitement and nervousness.

7.1. Skin Diseases

A. Fungal and Yeast Infections

1. Ringworm

This is caused by the dermatophytes *Trichophyton mentagrophytes* and *Microsporum* canis. *Trichophyton* is more commonly found on outdoor rabbits whilst *Microsporum* is found on house rabbits. It is a zoonosis, and rabbits can serve as a reservoir for human infection.

Clinical signs: Dry scaly lesions initially on the nose and muzzle. The lesions may spread across the face, round the base of the ears and to the feet. The lesions are generally pruritic. Infection of young rabbits occurs in the nest box, which may become contaminated with the fungal spores. During nursing the young are in direct contact with the skin and fur around the teats, and it is easily transferred to their mouth and muzzle.

Diagnosis: Only *Microsporum* will fluoresce under ultraviolet light (Wood's lamp). Diagnosis is by microscopy and culture on Sabouraud's medium.

Treatment: Griseofulvin given orally at a dose of 25 mg/kg twice daily for 4-6 weeks is effective. The tablets can be broken and given in a little strawberry jam. Griseofulvin should not be given during pregnancy as it is teratogenic. Washes that can be used in conjunction with griseofulvin include povidone -iodine, 1% triclosan and enilconazole at a 1:10 dilution. Clotrimazole cream can be used topically. The environment should be thoroughly disinfected. If a shed or building needs to be treated, a solution of enilconazole can be made into a spray and used at a rate of 50 mg/m^2 twice weekly for 5 months. Steps should be made to eliminate wild rats and mice from the environment as these are often carriers of the disease.

2. Candidiasis

Clinical signs: A moist dermatitis, especially under the chin and dewlap, caused by the yeast *Candida albicans*.

Diagnosis: Microscopy of skin scrapings.

Treatment: The area should be bathed daily with 1% chlorhexidine. Daily application of chlorhexidine in aqueous cream, or daily application of an antimycotic should clear the condition in 10-14 days. Natamycin or enilconazole used at a 1:50 dilution are suitable.

B. Ectoparasites

1. Fleas

Rabbits can be hosts for fleas. It is most important to stress the importance of mxyomatosis vaccination as fleas are the main vector of the virus. Outdoor rabbits that may have contact with wild rabbits may acquire *Spilopsyllus cuniculi* (the rabbit flea) and *Echidnophagis mymecobil* (the stickfast flea). House rabbits may harbour *Ctenocephalides felis felis*, and *Ctenocephalides canis* (the cat and dog fleas).

Treatment: Sprays containing pyrethrin can be used. For house rabbits a spot-on preparation containing 10% imidacloprid can be used.

2. Cheyletiellosis

Cheyletiellosis caused by the non-burrowing mite *Cheyletiella parasitovorax* is the commonest skin condition seen in rabbits. It appears most frequently in the milder warm weather of spring. It is thought to be a common commensal ectoparasite which may cause no problems in normal healthy rabbits, and in many cases infection is asymptomatic. Cheyletiellosis is a zoonotic disease, producing a transient pruritic rash in infected people. It can also be spread to cats and dogs, and this may become more significant as the numbers of house rabbits increase.

Clinical signs: Mild to severe seborrhoeic lesions, usually on the neck or dorsum. The lesions may spread ventrally. The condition is pruritic.

Diagnosis: The mites and skin debris can be picked up on adhesive cellulose tape, and examined under the microscope. The mites are relatively large, and have hook like accessory mouthparts.

Treatment: The life cycle of *Cheyletiella parasitovorax* is 5 weeks. Ivermectin can be given by subcutaneous injection at a dose of 300- 400 *mg*/kg. Injections can be repeated every 2 weeks, for three doses.

Ivermectin is also effective if used topically. The rabbit can be washed in 1% selenium sulphide. The rabbit should be dried with a hairdryer, and not towelled.

Alternatively the rabbit can be dipped in 0.01% amitraz which must be diluted 1:500 to achieve this concentration. Dipping should be repeated weekly for up to 6 weeks. The hutch should also be well cleaned, and can be sprayed with an environmental flea products.

3. Demodecosis

Infection caused by the mite *Demodex cuniculi* is rare, although it may be seen in immunosuppressed rabbits. The mite is probably present in very low numbers in most rabbits but may only multiply and cause clinical symptoms in the very young or immunosuppressed individuals.

Clinical signs: A moist erythematous dermatitis, with small pustules 2-4mm in diameter. It is not pruritic.

Diagnosis: Skin scraping. Several skin scrapings may be required to detect the mite. *Demodex cuniculi* is a long thin mite similar to the canine mite, when viewed by microscopy.

Treatment: Ivermectin by weekly subcutaneous injection. A dose of 500 *m*g/kg may be required to effect a cure. Alternatively the rabbit could be bathed weekly in a 0.01% solution of amitraz.

4. Ear mites

Infection with *Psoroptes cuniculi*, the ear mite, is mentioned here, as although it commonly affects the ear, it is capable of causing skin lesions as well.

Clinical signs: Early cases may have mildly pruritic ears, and rabbits can carry low grade infections for long periods of time. More advanced cases have thick yellowish grey crusts in the ears, which are very inflamed. The condition is pruritic. Rarely the lesions may spread across the face and limbs. Crusty lesions have also been reported on the ventral abdomen around the vent. In this site it must be differentiated from treponemiasis (rabbit syphilis).

Diagnosis: Microscopy. The mite will be present in large numbers in the lesions. They may be visible to the naked eye as they may reach 0.7mm in size. Examination under the microscope reveals their oval body shape, pointed mouthparts and three jointed pedicles with funnel shaped suckers.

Treatment: Ivermectin is given by subcutaneous injection at a dose of 400 *mg*/kg. The injection should be repeated on two more occasions at fortnightly intervals, as the mite takes 21 days to complete its life cycle. If the ears are to be cleaned the rabbit should be sedated, as removal of the crusts is very painful. The ears and skin can be bathed with dilute chlorhexidine, although there is no actual need to do this because once treatment has begun the crusts will dry up and fall out after a few days.

A short-acting corticosteroid can be given by injection to limit the inflammation. A few drops of 1% ivermectin can also be put directly into each ear. In-contact rabbits should also be treated, and the environment must be disinfected and sprayed with an acaricide, as the female mites can live off the host for several weeks making reinfestation possible.

5. Fur mite

Leporacarus gibbus (formerly *Listrophorus gibbus*) is another mite commonly found in the fur of rabbits. Its role in the development of dermatitis is uncertain, whether it is a commensal or a pathogen. Unlike *Cheyletiella* there are no reports of it as a zoonosis.

Clinical signs: Alopecia, pruritis and a moist dermatitis. The classic seborrhoea associated with Cheyletiella is not a feature. Many infestations may be asymptomatic. The eggs attach to the hair shafts, as do the hatched eggs, and empty larval and nymphal cuticles are left on the hair as the mites develop into adults. This debris gives the coat a 'salt and pepper' appearance.

Diagnosis: Microscopy. The adult female is a similar shape to a flea and much of the cuticle is finely striated. The head is sclerotised, giving it a dark appearance. The legs are short with no clasping adaptations; the mites attach to the hair using membraneous flaps arising from the coxae of the first pair of legs. The males have elongated adanal processes with adanal suckers.

Treatment: Ivermectin injections given weekly for 3 weeks at a dose of 400 mg/kg are effective. Alternatively, the rabbit can be shampooed in a 0.5% carbaryl shampoo weekly for 4 weeks. Dichlorvos strips can be hung in the shed as an environmental control.

6. Burrowing mites

Other mites that are occasionally found on rabbits are *Notoedres cati,* and *Sarcoptes* species. These mites are zoonotic.

Clinical signs: The lesions are spread over the head, especially around the base of the ears. Infestations by both mites are pruritic.

Diagnosis: Microscopy.

Treatment: Two injections of ivermectin at a dose of 400 *mg*/kg given a week apart are effective. Alternatively the lesions can be painted with 10% benzyl benzoate every 5 days.

7. Red fowl mite

Rabbits that are housed with chickens or other fowl may be affected by Dermanyssus gallinae which causes intense pruritis.

8. Ticks

Ticks are occasionally found on rabbits that live outdoors. They are removed in a similar fashion to those on cats and dogs, by applying methylated spirit or petroleum jelly to kill the tick, and then removing with a tick picker.

C. Bacterial Skin Diseases

1. Cutaneous abscesses

Abscesses are very common in rabbits; they may occur anywhere on the body where they are the result of any form of skin trauma, and commonly around the head and neck where they may be associated with dental problems. The nature of rabbit pus is very thick and tenacious. However, abscesses are generally not painful, and do not cause the rabbit to feel unwell; rabbits are often able to live with an abscess for months or years whilst on long-term antibiosis.

Clinical signs: A large non-painful swelling, which when aspirated reveals thick yellow purulent material. Microbiological culture often reveals Staphylococcus aureus or Pasteurella multocida. A culture should always be taken from the wall of the abscess as the centre of the purulent material is likely to be sterile. Infrequent complications are septicaemia, septic embolus and haematogenous spread.

Treatment: Isolated, well circumscribed abscesses are best treated by complete surgical excision, followed by antibiosis for 2 weeks post surgery.

2. Acute cellulitis

This syndrome is also associated with *Staphylococcus aureus* or *Pasteurella multocida.*

Clinical signs: An inflamed, painful oedematous area, usually around the head or neck. The rabbit is febrile, with a temperature of 40-42.2°C (104- 108°F). The lesion may develop into an abscess, or the skin over the area can become necrotic.

Treatment: Aggressive antibiosis. Analgesia should be given to reduce the temperature and manage the pain. Cool baths will also help bring the temperature down.

3. Blue fur (Moist dermatitis)

A moist dermatitis is seen associated with *Pseudomonas aeruginosa,* which affects any skin folds which are kept permanently wet. The skin may be kept wet by saliva or urine, or damp from leaking sipper bottles. The organism is thought to be harboured in drinking bottles, and these should be regularly cleaned.

Clinical signs: The fur classically takes on a blue-green discolouration particularly around the dewlap as the rabbit may 'chin' the water bottle.

Treatment: The area should be bathed and dried. Hair around the lesion should be clipped off. Antibiotics should be used topically, and given parenterally. Enrofloxacin and gentamicin have the best action against *Pseudomonas.* Any dental or urinary problems should be corrected, and cage sanitation improved.

4. Staphylococcosis

Staphylococcus aureus can cause a pustular dermatitis of young rabbits. Rabbits up to 10 days of age develop moist lesions on the ventral abdomen and medial aspect of the hind legs. Older rabbits (2-4 weeks of age) have numerous small abscesses over the body and a purulent conjunctivitis. Affected rabbits may develop a fatal septicaemia, with abscesses in the liver and other organs. In-contact lactating does may develop a suppurative mastitis, and suckling rabbits may also die as their dams develop agalactia.

Treatment: Antibiotics should be used topically and systemically. As this condition is associated with poor environmental conditions, hutch hygiene should be improved.

5. Necrobacillosis (Schmorl's disease)

This condition is caused by the bacterium *Fusobacterium necrophorum* and presents as necrotic lesions all over the body, most commonly around the head and lips. The rabbit may also develop multiple abscesses. Affected animals have general malaise and weight loss, and a common sequel is the development of pneumonia. The prognosis is poor.

Treatment: Antibiosis in the early stages. Surgical removal of the abscesses and necrotic tissue may be possible in some cases.

6. Rabbit syphilis

This disease is caused by *Treponema cuniculi*. The organism is transmitted by direct contact, via doe and buck during mating, and via the doe to her kits at parturition and during lactation. The rabbit spreads its lesions from its genitals to its face and legs as it grooms itself.

Clinical signs: Crusty lesions are found, primarily on the genitalia, but they also appear on the lips, nose, eyelids and perineum. Occasionally lesions may appear on the face and legs alone, whilst the genitalia remain seemingly unaffected. The lesions are not pruritic and begin as oedema, and then form vesicles which burst and become covered with a crust. The inguinal lymph nodes may be enlarged. Affected rabbits are not clinically ill. Other symptoms of reproductive disorders are infertility, stillbirths, neonatal mortality and metritis. The period between infection and the development of the sores is at least 8 weeks. Symptomless carriers can exist for months before the disease is triggered by stress.

Diagnosis: The condition must be differentiated from ringworm, and ectopic Psoroptes cuniculi infection. Deep skin scrapings will determine the bacterium. Silver staining of biopsies will also demonstrate the organism. Serology tests are also available to determine the presence of antibodies.

Treatment: The disease may be self-limiting, but recovered rabbits become symptomless carriers. For treatment procaine penicillin can be used at a dose of 40 000 IU/kg given by intramuscular injection once daily for 7 days. Alternatively an injection can be given once a week for 3 weeks at a dose of 80 000 IU/kg. Although penicillin is considered dangerous for rabbits because of the potential of developing enterotoxaemia, treatment of syphilis is the exception where it should be used. As long as the rabbit is on a good high fibre diet (hay and greens) with a probiotic, the effects of the antibiotic should be minimal. All affected, and in-contact rabbits should be treated at the same time.

D. Viral Skin Conditions

1. Myxomatosis

This is caused by the myxoma virus which is a pox virus.

Clinical signs :

Acute form: The rabbit develops oedematous swellings around the eyes, base of the ears and genitals. The swelling of the genitals is pathognomic for myxomatosis. There is a purulent blepharo-conjunctivitis which progresses to blindness. The rabbit may maintain its appetite initially, but will become anorexic as the disease progresses. Secondary Pasteurella pneumonia is very common and the usual cause of death.

Chronic (nodular) form: The rabbit develops oedematous swellings (pseudo-tumours), especially on the ears, nose and paws, 10-15 days after infection. These will spontaneously resolve, although the resultant scabs take longer to disappear.

2. Warts

Wart-like lesions are caused by the *Shope papillomavirus*. The warts are usually found on the eyelids and ears. The virus is transmitted from wild rabbits by arthropod vectors. Although these warts are generally benign they may be the precursors to carcinomas.

E. Miscellaneous Skin Conditions

1. Moulting

Rabbits may moult every 3 months. The moults usually alternate between light and heavy shedding. It is very common in Angoras and Cashmeres, and some may go completely bald. Hay must be fed at all times to prevent excessive hair ingestion and to ensure swift passage of any hair through the digestive system. Any loose hair should be removed from the cage every day to limit the amount ingested. Some rabbits may act as if they are very depressed during their moult, and they may become anorexic. They should be tempted with fresh food, in particular chicory, comfrey and spinach. A probiotic can also be given, and this may help stimulate the appetite. If a rabbit appears to be stuck in moult it is useful to feed some groundsel, as this will often help it to complete the moult. A few drops of cod liver oil can be added to the dry food twice a week. A tonic containing vitamin B1 and minerals can be added to the drinking water at a dilution of 20 ml/litre.

2. Alopecia

(1) Hair loss on the chest and abdomen is common in does during pregnancy and pseudopregnancy, as they pluck their hair to build nests. Some does can make their skin very sore. If pseudopregnancy is a recurrent problem then ovariohysterectomy is curative. Rarely, nest building has been reported in neutered males.

(2) Hair loss may also be due to barbering. Some rabbits may barber themselves, in which case the hair loss is from the body. Rabbits that are being barbered may have hair loss around the head and dewlap. In cases where a dominant rabbit is barbering its companions, the rabbits may need to be segregated. Increasing the fibre content of the diet should reduce barbering, as it will reduce boredom, and provide a constant source of material to chew. Similarly, toys can be introduced to reduce boredom. A single rabbit that is self-barbering will benefit from neutering and pairing with a companion. Reducing the light intensity and reducing day length will reduce barbering.

(3) Nutritional magnesium deficiency will cause hair loss. To correct this deficiency magnesium oxide can be added to the diet at 0.25%.

3. Atopy

Clinical signs: Alopecia, pruritis, crusting, erythema, self-trauma of the ears, nose, footpads, thighs, medial hocks and abdomen. There may be accompanying conjunctivitis and epiphora, or epiphora may be the only clinical sign.

Treatment: Parental corticosteroids, and steroid eye drops. The antihistamine diphenhydramine can also be used.

4. Autoimmune skin disease

Facial skin lesions similar to pemphigus have been reported. Lesions predominate around the face, nose, lips and chin. A response is seen to corticosteroid treatment. Prednisolone can be given at an initial dose of 2 mg/kg daily; if a response is seen this dose may be reduced.

5. Fly strike (blowfly myiasis)

Blowflies are attracted to any damp, soiled areas, and will lay their eggs; in warm, damp conditions these will rapidly hatch into larvae (maggots) in 24 hours. A prime site is the perineum, particularly in overweight rabbits which may have a pouch of skin which remains constantly soiled. Similarly at risk are elderly or disabled rabbits that are unable to clean themselves properly. The larvae burrow under the skin, secreting a chemical similar to local anaesthetic as they go, and the rabbit may be unaware of their presence. The affected rabbit will rapidly become quiet and depressed. The damage to the tissues by the larvae causes hair matting and a moist dermatitis.

Treatment: Initial treatment is aimed at removing the larvae. The rabbit should be sedated if necessary. The area can be sprayed with diluted ivermectin. The maggots are picked off with forceps. A single injection of ivermectin at a dose of 400 mg/kg is given subcutaneously. The lesions can be bathed with 1% chlorhexidine, or a 10% povidone iodine solution. The skin deficits can be treated with Dermisol cream. The rabbit should also be treated for shock. Fluids can be given subcutaneously. Analgesia must be given. Flunixin meglumine can be used at a dose of 1.1 mg/kg by intramuscular injection. Antibiotics are given, either oxytetracycline at a dose of 30 mg/kg by injection every 3 days or enrofloxacin at a dose of 10 mg/kg orally daily. For severely affected rabbits euthanasia may be the only option.

Prevention: Owners must be made aware of the importance of checking their rabbit's vents daily for signs of soiling, and for the presence of fly larvae. The rabbit's diet should be improved to contain higher fibre (hay and plant fibre) to reduce the incidence of 'sticky bottom syndrome'. Rabbits with a pouch of skin around the perineum may benefit from a perineal skin fold reduction. If the rabbit's vent area needs to be cleaned regularly it is better to use a skin cleaner which will soften the faeces, rather than get the whole bottom of the rabbit wet. Shaving the area regularly will help keep the area from becoming soiled. Attention to cage hygiene should be paramount. Herbal sprays containing citronella, a natural fly repellent, may be of value, or a pyrethrin-based flea spray can be used around the vent. Other measures to reduce flies in the rabbit's environment should be taken; the use of dichlorvos strips in the shed, or the spraying of the hutches with a household flea spray.

6. Cuterebra (Rabbit warble fly)

In some countries the rabbit warble fly Cuterebra will cause a remarkable form of myiasis. A single larva can pupate in the subcutaneous tissue of a rabbit, and become encapsulated in a large fistulous lesion.

Clinical signs: A swollen mass up to 3 cm in diameter, often around the neck or rump. The swelling will have a single air hole, and the larva can be seen moving inside.

Treatment: Surgical removal of the larva through the air hole. Antibiotics should be given to prevent secondary infection.

7. Pododermatitis (Sore hocks)

Clinical signs: Hair loss, scaling, erythema and ulceration of the skin on the plantar aspect of the metatarsus. Very occasionally the front feet may be affected too. The development of this condition is associated with abrasive or soiled cage floors, in conjunction with obesity and in rabbits that stamp frequently. In house rabbits it may be associated with **'carpet-burn'**. It is common in rabbits that only have a thin covering of fur on the foot pad, especially Rex rabbits. The lesions become secondarily infected with Pasteurella multocida or Staphylococcus aureus, and may develop caseous abscesses. Occasionally the condition may progress to cause osteomyelitis and septicaemia.

Treatment: The lesions should be cleaned with an antiseptic such as povidone iodine. If the rabbit will tolerate dressings, the feet may be dressed. Dressings should be changed every 5 days, and once the hock is healing the area can be protected with a lighter covering such as a small sock. A soft, clean surface, either deep dry bedding or towels, should be provided. The injection of a long-acting steroid under the lesion may be of benefit. The lesions can take weeks to heal. A tendency towards sore hocks may be inherited, and affected animals should not be bred from. Skin trauma associated with rabbit rings Breeders use rings on the hind leg of rabbits for identification purposes. These rings are slipped up the leg and over the hock when the rabbit is young, and as the rabbit matures the ring is unable to slip off. When show rabbits are kept as pets they may put on weight and the leg band can become tight. If unnoticed the band can cut through the skin which exposes the soft tissue structures underneath. In advanced cases there may be swelling of the distal limb.

Treatment: The ring should be removed, and there are special clippers available to enable this to be done in the least traumatic way possible. If the ring is deeply embedded the rabbit may need to be sedated first. The area of damaged tissue should be cleaned with an antiseptic regularly. Care must be taken to avoid **'fly strike'** whilst the tissues are healing. It is recommended that any rabbit that is not shown should have its ring removed.

8. Skin tearing

A connective tissue defect similar to the Ehlers - Danlos syndrome has been reported in rabbits.

9. Perineal skin fold dermatitis

This condition occurs mostly in older does. The excess skin in the perineal region causes a flap which becomes soiled with faeces (caecotrophs) and urine, leading to the development of a moist dermatitis. This soiling leads to an increased risk of 'fly strike'. Some does will develop anorexia and pyrexia.

Treatment: The area can be gently bathed and cleaned daily with a dilute solution of chlorhexidine. The dermatitis can be treated with an antibiotic corticosteroid cream. The skin can be protected with petroleum jelly, zinc and castor oil cream or talcum powder. Severe cases may benefit from surgery to resect the skin folds. Procedure for perineal fold reduction: Before surgery it is important to correct the diet to prevent excess caecotroph production and reduce obesity. Any skin infection at the site of surgery must be treated before performing the surgery. The rabbit is placed in dorsal recumbency, and the skin of the ventral abdomen is clipped and prepared for surgery. A crescent of skin is removed cranial to the genital area, and some of the subcutaneous fat can also be removed. The amount of skin resected should be just enough so that once sutured the ventral skin surface is flat, but not under tension. The incision can be sutured in two layers, an absorbable layer of sutures in the subcutaneous fat to take the tension, and simple interrupted sutures in the skin layer.

10. Neoplasia

Skin neoplasms include lipomas, squamous cell carcinomas, trichoepitheliomas and basal cell tumours. All of these are slow to metastasise, and surgical excision is curative.

7.2. Diseases of Reproductive System

A. Complications of Pregnancy

1. Pregnancy toxaemia (Ketosis)

Pregnancy toxaemia is rare in rabbits. When seen, this condition is most common in does carrying their first litter. It occurs during the last week of pregnancy, and is most frequently seen in does that are overweight. It has also been seen in pseudopregnant and post-partum rabbits. If the doe becomes anorexic, body fats are broken down to provide energy and the subsequent ketone bodies enter the bloodstream. Fatty liver (hepatic lipidosis) develops. If the doe aborts or gives birth in the early stages the condition may resolve, but if the condition progresses the prognosis is guarded.

Clinical signs: Lethargy, dullness of the eyes, salivation, respiratory distress, convulsions and collapse. The breath may have the classic acetone smell of ketosis. The urine becomes acidic (pH 5-6).

Treatment: Glucose should be given orally, or as a subcutaneous or intraperitoneal injection of 5% glucose. A short-acting corticosteroid injection may be beneficial. The rabbit can be force-fed baby cereal and fruit puree.

Prevention: Young does should not become over fat before breeding. Body fat mass can be reduced by decreasing the concentrate feed, and introducing more roughage in the form of hay and straw. Exercise is important. In a small hutch even placing the food and water sources at opposite ends of the hutch will increase exercise. Any stress which might trigger anorexia in late pregnancy must be avoided. An electrolyte solution or probiotic containing glucose can be put in the drinking water during the last week of pregnancy for susceptible does.

2. Pseudopregnancy

Pseudopregnancy may occur after an unsuccessful mating, or following ovulation triggered by the mounting of another doe. This condition usually lasts for about 18 days, and is accompanied by fur plucking, nest building and mammary gland development. The doe may also become aggressive and territorial. If milk is produced, care must be taken that she does not develop mastitis. During this time the doe is unresponsive to being re-mated. Some does that live alone may be able to self-induce ovulation and can exhibit recurrent false pregnancies; for these ovariohysterectomy is curative, or for the short term a hormone injection can be given. Delvosterone containing 100 mg/ml proligestone can be given at a dose of 30 mg/kg. Hydrometra is a rare condition which is thought to arise from a pseudopregnancy which does not reverse, and under the continued influence of progesterone the endometrial glands continue to secrete fluid which accumulates in the uterus.

3. Miscarriage and abortion

Foetal death in the first 3 weeks of gestation will result in resorption, in the last week abortion occurs. There are many possible causes.

(1) **Nutrition.** Obese does may be prone to resorption, or the development of ketosis and subsequent abortion. Poor nutrition and low energy intake of the doe is associated with foetal resorption, small litter size and the delivery of weak or dead kits. Deficiencies of vitamin A, E or protein deficiency will result in the loss of litters.

(2) **Infections.** Stillbirths and neonatal mortality can be associated with syphilis (*Treponema cuniculi*). Late term abortion can be caused by *Listeria* infection. *Pasteurella multocida* can cause epididymitis and orchitis in the male, and metritis in the female, and hence reduced fertility of both sexes.

(3) **Other causes**. Stress or trauma. Some rabbits may have a genetic predisposition towards litter loss.

B. Problems of Parturition

1. Cannibalism

Cannibalism of the young may occur. This may be accidental, inexperience (first time does), or in response to environmental pressures such as overcrowding. Cannibalism may occur by other does, reverting to their natural survival instincts to cannibalise kits that do not belong to them, to ensure survival of their own line.

2. Dystocia

Dystocia is uncommon, and generally associated with foetal oversize or malpresentation. Obesity or malnutrition may also predispose to dystocia. Oxytocin can be given if parturition is delayed, at a dose of 1-2 IU/kg by intramuscular injection. Calcium gluconate can be given orally at a dose of 5-10 ml of a 10% solution if uterine inertia is suspected. Caesarian section or ovariohysterectomy can be performed on valuable animals.

3. Uterine prolapse

This is an uncommon condition, but has been recorded associated with dystocia. The prolapsed uterus rapidly becomes engorged with blood, and contaminated with foreign material. Ovariohysterectomy, after first pulling the uterus back into the abdomen, is curative.

C. Conditions of the Uterus

1. Endometrial hyperplasia

The endometrium undergoes changes as the rabbit ages, and the endometrial glands become cystic and hyperplastic. The mammary glands may also become cystic.

Clinical signs: Intermittent haematuria, with the blood being passed at the end of urination. Blood clots may appear with the urine. The rabbit becomes lethargic. The thickened womb may be evident on abdominal palpation, or by radiography. In a breeding doe the first signs may be reduced fertility.

Treatment: Ovariohysterectomy. If necessary histopathology can be performed to differentiate between hyperplasia and adenocarcinoma. Cystic ovaries are commonly associated with this condition.

2. Uterine adenocarcinoma

This is the most common condition seen in unneutered does. It may occur in rabbits from 2 years of age, and 60 - 80% of does may be affected by the age of 6 years. It is the principal reason why does should be neutered before 2 years of age. It is thought that falling oestrogen levels in the older doe may be responsible for the development of neoplasia.

Clinical signs: The early stages of the disease are asymptomatic, and many uterine growths are first found during routine hysterectomy in the older doe. As the disease progresses, intermittent haematuria is a feature. The blood associated with uterine pathology generally occurs at the end of urination, rather than mixed evenly with the urine as with diseases of the urinary tract. The presence of blood clots also suggests uterine pathology. The doe may also become more aggressive, or develop anorexia. The growth is often palpable on abdominal examination, or can be confirmed by radiography. The mammary glands may also become cystic. Metastases are uncommon, but there may be spread to the skin, lungs or liver. In a breeding doe the first signs may be those of reduced fertility, and foetal loss or resorption.

Treatment: Ovariohysterectomy. As with routine neutering, the uterus should be ligated caudal to the double cervices to ensure the removal of all uterine tissue. A chest radiograph should be taken before surgery to detect metastasis, particularly if the doe is suspected to have an advanced stage of the disease. The prognosis for does with no signs of metastasis is good following ovariohysterectomy.

3. Endometrial venous aneurysm

This condition may mimic the clinical signs of uterine adenocarcinoma. The aetiology of this condition is unknown; some does develop large bloodfilled uterine horns.

Clinical signs: These are similar to urinary tract infection; the doe may pass blood during or after urination. In the later stages large quantities of blood may be passed. The enlarged womb may be evident on abdominal palpation, or diagnosed by radiography.

Treatment: Ovariohysterectomy.

4. Hydrometra

This is an uncommon condition, associated with the accumulation of watery fluid within the womb. The aetiology of this condition is unclear, but it is thought to be associated with self-ovulation and pseudopregnancy in the doe. The increase in progesterone stimulates secretory activity of the endometrial cells, and closure of the cervix. If these changes are not reversed at the end of pseudopregnancy, hydrometra can develop.

Clinical signs: Cases may not be presented until the condition is advanced. The uterus may swell until it occupies 80% of the abdominal cavity; hence the rabbit may look normal or obese, whilst actually losing body condition. Eventually the doe becomes anorexic, as the womb presses on the digestive system. Diagnosis is by abdominal palpation and radiography.

Treatment: Ovariohysterectomy.

5. Endometritis and pyometra

This condition can occur both in pet (unbred) rabbits and breeding does. In the latter it can occur following a recent mating, or following parturition or a pseudopregnancy. Unneutered does housed with castrated bucks that occasionally mount them may have an increased incidence of pyometra. Mild cases of endometritis may show no clinical signs, and still be able to breed. Pasteurella multocida and Staphylococcus aureus are most commonly isolated from cases of pyometra.

Clinical Signs: Cases of pyometra develop anorexia, depression and a creamy vaginal discharge. Haematology may reveal a raised white cell count with a neutrophilia. On abdominal palpation the uterus may feel 'doughy', and will be enlarged on radiography.

Treatment: Ovariohysterectomy, which may be complicated by the presence of adhesions to other organs. If there is a concurrent peritonitis the prognosis is very poor. Fluid therapy is important, and antibiotic therapy should be commenced, preferably based on the results of a bacterial culture and sensitivity.

D. Conditions of the Mammary Glands

1. Mastitis

Mastitis may occur in lactating does that are kept in unsanitary conditions. Teat trauma, injury to the teats, or biting of the nipples will also predispose to the development of mastitis. Mastitis can also occur in association with pseudopregnancy. The causal bacteria are usually *Pasteurella, Staphylococcus* and *Streptococcus* spp.

Clinical signs: One or more of the mammary glands will be hot and painful. The affected doe may be pyrexic, anorexic and reject her young. The condition may be fatal for both the doe and her young. Death of the doe is associated with septicaemia.

Treatment: Antibiosis, preferably based on the results of bacterial culture and sensitivity of the milk from affected glands. The best antibiotics are generally enrofloxacin or a trimethoprim-sulpha combination. Analgesia is most important, and the glands can have warm compresses applied several times daily. The kits should be removed and the doe should be placed in a warm, clean environment. The kits should be hand-reared and not fostered, as fostering may spread the infection to the healthy foster mother.

2. Cystic mastitis

This is a condition of older does, over the age of 3 years. It occurs in nonbreeding does. It may be associated with uterine hyperplasia or uterine adenocarcinoma.

Clinical signs: One or more glands may be affected. The gland is swollen and bluish in colour. The gland may have a brown serosanguinous discharge from the nipple. The swelling is non-painful. The doe may show other symptoms generally associated with pseudopregnancy such as nestbuilding and aggression.

Treatment: Ovariohysterectomy. The affected gland can be removed at the same time, although this may not be necessary as the cysts may regress 3-4 weeks after surgery.

3. Neoplasia

Both mammary papillomas and adenocarcinomas have been recorded. The latter may be present in association with uterine adenocarcinoma. Radiographs should be taken to identify any pulmonary metastases before considering surgery. If the doe is entire, ovariohysterectomy should be advised at the same time.

E. Conditions of the Testicles

1. Epididymitis and orchitis

Infection with Pasteurella multocida can cause inflammation and abscessation of one or both testicles. Treatment with antibiotics is unrewarding, and castration should be considered in non-breeding males.

2. Traumatic wounds

When two buck rabbits fight, they will often attack each other's testicles. Depending upon the degree of trauma caused, some bucks can be sufficiently damaged to require castration. It is often thought that it is the dominant buck that inflicts the wounds; however, more commonly the dominant buck mounts the submissive buck backwards, and it is the buck underneath that attacks the uppermost buck in self-defence, rather than in aggression. For this reason it is generally advised that bucks are not kept together.

3. Neoplasia

These are conditions of older bucks. Both testicular seminomas and testicular interstitial cell carcinomas have been recorded. Metastases are unlikely and castration is curative.

F. Rabbit Syphilis (Vent Disease)

This disease is caused by *Treponema cuniculi.* The organism is transmitted by direct contact, via doe and buck during mating, and via the doe to her kits at parturition and during lactation. It then spreads from the genitals to the face of an infected rabbit by autoinfection.

Clinical signs: Crusty lesions, primarily on the genitalia, but these also appear on the lips, nose, ears, eyelids and perineum. The lesions are not pruritic. The lesions begin as oedema, and then form vesicles which burst and become covered with a crust. Affected rabbits are not clinically ill.

Reproductive disorders include male infertility, associated with prepucial inflammation. Does develop infertility, metritis, retained afterbirths, stillbirths and neonatal mortality. The period between infection and the development of the sores is at least 8 weeks.

Diagnosis: The condition must be differentiated from ringworm, and ectopic Psoroptes cuniculi infection. Deep scrapings will determine the bacterium. Silver staining of biopsy material will reveal the organism. Serology tests are also available to determine the presence of antibodies.

Treatment: Procaine penicillin at a dose of 40 000 IU/kg given by intramuscular injection once daily for 7 days. Although penicillin is considered dangerous for rabbits because of the potential of developing enterotoxaemia, treatment of syphilis is the exception where it should be used. As long as the rabbit is on a good high fibre diet (hay and greens) with a probiotic, the effects of the antibiotic should be minimal. All affected and in-contact rabbits should be treated at the same time. Alternatively 80 000 IU/kg can be given by injection once a week for 3 weeks.

7.3. Diseases of Urinary System

This is a common and normal finding in the rabbit. The rabbit is able to excrete porphyrins into the urine which will colour the urine red. It can be distinguished from true haematuria by a dipstick test which will confirm that there is no blood in the urine. Porphyrin excretion in the urine may increase in times of stress. The urine may also become discoloured if the diet is high in carrots or other sources of carotene, as the pigments are not completely broken down by the kidneys and are excreted in the urine. This may colour the urine orange or brown. Eating acorns will similarly lead to the production of red urine. Dehydration will increase the depth of colour, and may occur if water is not readily available, or if it is presented in a container with which the rabbit is unfamiliar. Illness may also lead to a decreased water intake and subsequent concentration of urine.

2. Haematuria

Haematuria must be first differentiated from 'red urine' by a dipstick test. The differential diagnoses are cystitis, pyelonephritis, urolithiasis, and in the intact female, reproductive disease such as uterine adenoma, adenocarcinoma or endometrial venous aneurysm. The blood associated with urinary tract disease is usually spread evenly in the urine, whereas the blood from reproductive disorders is often produced at the end of, or independently of urination, and may contain blood clots.

3. Cystitis

Clinical signs: Haematuria, dysuria and pain on urination. There may be perineal soiling.

Diagnosis: This condition must be differentiated from urolithiasis by radiography. Radiography may also indicate the presence of reproductive disorders in the intact female, which do appear to be more common than cystitis. A urine sample should be collected for urinalysis. This can be done by collection into a litter tray, by manual expression, cystocentesis or via catheterisation. Care must be taken when manually palpating the bladder as the bladder wall can be easily bruised or ruptured. *E. coli* and *Pseudomonas* are most commonly isolated in cases of cystitis.

Treatment: Antibiotics, based on urine culture and sensitivity, may need to be given for up to 4 weeks. Drinking should be encouraged, and the rabbit's fluid intake can be increased by offering plenty of fresh vegetables. Dandelions and cleavers are useful feedstuffs as they have diuretic properties. Vitamin C can be given at a dose of 50-100 mg/kg to help heal the damaged bladder. Cranberry tablets, or unsweetened cranberry juice can be used to prevent re-occurrence as the cranberry prevents bacteria sticking to the bladder wall.

4. Sludgy urine

Because the urine is naturally alkaline, triple phosphate and calcium carbonate crystals readily precipitate out, giving rabbit urine a cloudy appearance. However, if the urine becomes too full of sediment this may lead to urine that is so full of crystals that it is of toothpaste consistency, or ultimately lead to urolithiasis (bladder and kidney stones). Diets that are high in protein and calcium predispose to crystal development. The rabbit has an unusual calcium metabolism. It appears to have a higher plasma calcium level than other species, and there seems to be a more direct link between dietary calcium and plasma calcium. The plasma levels are controlled by increased excretion in the urine, rather than other homeostatic mechanisms. Unfortunately urinary acidifiers (used in other species to lower the pH of the urine and aid crystal dissolution) cannot be used in rabbits as they harm the digestive system by altering the intestinal flora. The exception to this is ascorbic acid (vitamin C) which can be added to the drinking water at a dilution of 200 mg/litre.

Clinical signs: Polyuria, dysuria, perineal soiling. Also anorexia and depression. Urinalysis may reveal E. coli or Pseudomonas infection, and this should be treated with an appropriate antibiotic. This condition can be corrected by altering the diet to a low protein and low calcium diet. Rabbits with kidney dysfunction also require phosphate restriction. Obese rabbits should be encouraged to exercise.

Treatment: Some cases may respond to dietary alterations. Antibiosis may be indicated to control secondary infection. Anorexic rabbits will require fluid therapy and analgesia. It is possible to catheterise bucks under sedation in order to flush the 'sludge' from the bladder. If catheterisation is not possible a cystotomy may be necessary to flush the bladder. Dietary control for calculi prevention This diet is appropriate for the correction of 'sludgy bladder', and also the long-term management of cases after bladder stone removal. The aim is to produce a diet low in protein and calcium and high in fibre.

(1) There should be unlimited access to good quality grass hay at all times, as the major fibre provider. Only grass hay should be used; alfalfa hay or blocks must not be given as they are high in calcium.

(2) Dry or pelleted food (which is high in protein and carbohydrate) should be restricted. The rabbit mix should not contain alfalfa.

(3) Processed 'treats' (full of carbohydrate and sugar) should be cut out and replaced with fruit treats.

(4) Vegetables and fruit can be slowly introduced from the following list, one at a time. If any item upsets the rabbit it

will show diarrhoea 24- 48 hours after its introduction, and that fruit or vegetable should be removed from the diet. Once the new regime is fully introduced the rabbit should receive at least half a cup of vegetables per kilogram of body weight daily.

(5) Astringent plants such as blackberry or raspberry leaves can be introduced immediately: they are a good source of fibre and will not cause diarrhoea.

(6) Any vitamin and mineral supplementation should be stopped. Drinking should be encouraged at all times. If necessary a small amount of sweet tasting fruit juice can be added to the drinking water to increase the rabbit's water uptake.

5. Urolithiasis

Clinical signs: The rabbit may strain to urinate, have genuine haematuria (not 'red urine'), and be stained or wet around the genitals. Pain may be exhibited as teeth grinding, an increased respiration rate and a hunched posture. The rabbit may also be anorexic and lethargic. Litter trained rabbits may lose their litter habits.

Diagnosis: The stone, or stones may be palpable on examination. Radiographs will show the stone and its position; as the uroliths contain primarily calcium they are radiodense. Uroliths are generally calcium oxalate, calcium carbonate, or less frequently ammonium phosphate. Some rabbits may have small stones in the kidneys, and sometimes stones will leave the bladder and become stuck in the urethra. Radiologically the stone will appear as a discrete radiodense mass, in comparison to 'sludgy bladder' when the whole bladder may be filled with radiodense material. Hydronephrosis or hydroureter may occur secondarily, and blood biochemistry may reveal a raised blood urea level.

Treatment: Bladder stones need to be removed surgically, as do stones lodged in the urethra. Stones lodged in the urethra should be retropulsed into the bladder for removal from there. A midline incision is made over the caudal abdomen. The bladder is reflected through the incision and the cystotomy is performed through the dorsal surface of the bladder awayfrom any blood vessels. The bladder should be closed in two layers of simple interrupted sutures with an absorbable suture material such as vicryl. Post-operative analgesia should be given, and continued for 24-72 hours, until the rabbit is urinating normally. Fluids should be given pre- and post-operatively. Antibiosis is needed, preferably based on the results of a urine culture, and may be required for several weeks.

Prevention: The diet needs to be gradually altered to one of lower calcium content. Fluid intake should be increased, if necessary by adding some sweetened fruit juice to the drinking water, this will help the rabbit diurese itself and keep the urine dilute. Chewable cranberry tablets or cranberry tablets with vitamin C can also be given. The cranberry contains natural ingredients which prevent bacteria adhering to the bladder wall, and the vitamin C helps repair the damaged bladder. These are used in preference to cranberry juice which is often too high in sugar content. The rabbit should be encouraged to exercise, and put on a restricted energy (hay and greens) diet if it is obese. Wild plants such cleavers have a natural diuretic action, and yarrow is of benefit as a urinary antiseptic. Dandelions are also diuretic, but should be given in moderation as they are a source of calcium.

6. Perineal urine scalding

Although this is most commonly associated with urinary tract disorders, it may occur in association with any condition that makes the rabbit reluctant to move. Spinal pain, pododermatitis, paraplegia, obesity and concurrent illness may all make the rabbit urinate where it is sitting and cause urine scalding. Subsequently the perineum becomes damp and sore. The underlying cause should be investigated and treated. The area should be cleaned daily with a gentle antiseptic such as chlorhexidine and waterresistant antiseptic cream such as zinc and castor oil cream can be applied to the perineum.

7. Urinary incontinence

Urinary incontinence may occur in female rabbits after ovariohysterectomy. This incontinence is responsive to diethystilbestrol at a dose of 0.5 mg given orally 1-2 times weekly. Phenylpropanolamine can also be used at a dose of 6.25-12 mg/rabbit twice a day. Side effects may be inappetence or hyperactivity. An atonic bladder can be treated with bethanechol, as this increases smooth muscle tone. The dose is 2.5-5 mg orally twice a day. Incontinence may also be associated with lumbosacral vertebral fractures and dislocations, or central nervous system lesions caused by *Encephalitozoon cuniculi*. Often the loss of litter-box training may be the first sign of arthritis or paresis in the older rabbit. The resultant perineal scalding should be bathed daily with a mild chlorhexidine solution, and a topical antiseptic cream applied.

Ectopic ureters are possible, but rare. Urolithiasis, cystitis and 'sludgy bladder' will also present as incontinence and perineal soiling.

8. Urinary catheterisation

Catheterisation is useful for obtaining urine samples, for flushing the bladder of 'sludge', or for contrast radiography. Tom cat catheters may damage the urethral wall, and a softer 3-6 Fr (1-2mm diameter) flexible catheter or feeding tube should be used. Analgesics should be given after catheter removal. If an indwelling catheter is placed, the rabbit should wear an Elizabethan collar to prevent it chewing at the catheter.

9. Glycosuria

Glucose may be identified in the urine in times of stress, ketosis and hepatic lipidosis. The latter is possible in any rabbit that has not eaten properly for 48 hours. Diabetes is rare in the rabbit. The urine should be rechecked on a sample taken at home to rule out stress. Cases of ketosis will have an accompanying ketonuria. A blood sample to check liver enzymes will be necessary to investigate hepatic lipidosis.

10. Acute renal failure

Clinical signs: Lethargy, depression, anorexia, polyuria, polydipsia and perineal urine scalding.

Diagnosis: Biochemistry. Urinalysis may reveal proteinuria, haematuria and pyuria. The urine should be cultured to determine if the cause is infectious. *Pasteurella multocida* and *Staphylococcus* are often the causal agents of pyelonephritis.

Treatment: Fluids must be given intravenously to promote diuresis, at a rate of 100 ml/kg per day. If an infectious agent is identified the appropriate antibiotics should be given. Enrofloxacin is a useful antibiotic. Trimethoprim-sulpha combinations and aminoglycocides should be avoided as they are potentially nephrotoxic. The prognosis is guarded.

11. Chronic renal failure

This may occur associated with hypercalcaemia and renal calcinosis. Interstitial fibrosis and chronic interstitial nephritis can occur in older rabbits, and fatty degeneration of the kidneys may be seen in obese rabbits.

Clinical signs: Lethargy, depression, anorexia, polydipsia, polyuria and perineal urine scalding. Anaemia is a common finding.

Diagnosis: Biochemistry will reveal a raised BUN and creatinine. Urinalysis may show a proteinuria. Haematuria and pyuria are less frequent findings. A glucosuria may also occur with moderate to severe tubular damage. Radiography may demonstrate calcium crystals in the kidneys.

Treatment: Supportive treatment to promote diuresis. Where high blood calcium and renal stones are identified, the diet should be changed as for urolithiasis. Anabolic steroids can be given to reduce the uraemia and improve the appetite; B vitamins should also be given. The prognosis is guarded.

12. Renal neoplasia

Neoplasia may be identified at post-mortem. Tumours include renal adenomas, adenocarcinomas, nephromas and lymphomas. Embryoma nephromas are the third most common neoplasm in the rabbit, and sometimes can be felt by abdominal palpation.

13. Renal cysts

Subcapsular cysts can develop in the kidney. They are thought to be an inherited condition. They are of no clinical significance, but an enlarged cystic kidney may be evident on abdominal palpation.

14. Encephalitozoonosis

This is a common disease caused by the protozoan Encephalitozoon cuniculi. The protozoa have a predilection for renal and central nervous tissue. The spores multiply in both sites, and are shed through the urine. The presence of the organism does not necessarily cause disease.

Clinical signs: These generally reflect the central nervous system infection, rather than renal infection. Urinary incontinence may occur as a result of the CNS involvement.

Diagnosis: Antibody titres can be measured. However, it is impossible to distinguish between an active infection, a previous infection or an asymptomatic infection. A negative titre does, however, rule it out of the differential diagnosis. At post-mortem the kidneys are pitted and scarred with multiple white spots on their surface.

7.4. Diseases of Respiratory System

Respiratory disease is a major cause of morbidity and mortality in rabbits. Respiratory disease in the rabbit is generally bacterial in origin, rather than viral. *Pasteurella* is the most frequently isolated bacteria, but disease may also be associated with *Bordetella bronchiseptica, Staphylococcus aureus, Moraxella catarrhalis, Mycobacterium, Pseudomonas aeruginosa* and *Mycoplasma*. Rabbits are obligate nose breathers, and respiratory distress can occur readily in cases of upper or lower respiratory tract infection. Mouth breathing indicates severe distress and carries a guarded prognosis. Dyspnoea may be a symptom of lower respiratory infection, but it may also accompany any cause of abdominal pain, in particular urinary calculi or uterine adenocarcinoma. The patient's respiration should be studied before it is handled and stressed. The normal respiratory rate is 30-60 breaths per minute and is accompanied by nose twitching.

Trigger Factors in Disease

There are many factors that can predispose to respiratory disease, and these should be taken into account when investigating a sick rabbit. Nutrition is important, as a hay and fresh vegetable diet can protect against the development of disease. The presence of ammonia either due to poor sanitation, or the feeding of high protein diets, will weaken the respiratory mucosa. Sudden changes in environmental temperature or draughts can trigger infection, and outbreaks are most common in spring and autumn when the temperatures are most variable. The actual environmental temperature is not that important, it is sudden fluctuations that must be avoided. Stress of any sort (age, pregnancy, shows, etc.) is another trigger factor. In house rabbits the presence of air conditioning or central heating can lower the humidity and increase the amount of airborne dust particles. This may weaken the respiratory mucosa and predispose it to disease. The house rabbit may also be exposed to cigarette smoke which can damage the mucosa.

A. Bacterial Diseases

1. Pasteurellosis

Pasteurellosis is caused by the bacteria Pasteurella multocida. The disease caused depends upon the virulence of the strain involved, and the resistance of the host. The classical clinical signs of rhinitis ('snuffles') are not seen in rabbits under 12 weeks of age due to the presence of maternal antibodies for the first 8 weeks in endemic groups, and due to the fact that the nasal sinuses of the young rabbit are insufficiently developed to allow colonisation of the bacteria. The simultaneous presence of Bordetella bronchiseptica in the sinuses facilitates Pasteurella colonisation. Not all rabbits that carry Pasteurella become ill. Some may spontaneously eliminate the infection, whilst others may become chronic carriers. Healthy rabbits housed with infected rabbits may not get the disease if their resistance is good. Pasteurellosis tends to be a disease of intensively housed rabbits, and is less of a problem in house rabbits.

Clinical signs: The classical symptoms are rhinitis, conjunctivitis, sinusitis and dacrocystitis - the complex referred to as **'snuffles'**. The affected rabbit has a white discharge from its nose, and the fur on the medial aspect of the forelegs will be matted and sticky where it wipes its nose.

Other symptoms are otitis media and otitis interna (causing 'head tilt', nystagmus and ataxia), pleuropneumonia, pericarditis and tooth root abscesses. Symptoms of lower respiratory tract infection are anorexia, weight loss, depression and fatigue, with dyspnoea on exertion. These symptoms can occur in the absence of any upper respiratory tract infection. The prognosis for dyspnoeic rabbits is poor.

Abscesses can also form in distant sites causing testicular abscesses, endometritis, and pyometra. Septicaemia caused by Pasteurella is one of the commonest causes of sudden death, which may occur following stress, with few preceding clinical signs.

Transmission: The most common form of transmission is by direct contact, and by inhalation of airborne droplets containing bacteria. The bacteria can survive for several days in moist secretions and water. Indirect spread can occur via bowls and drinkers contaminated by nasal secretions. Rabbits can only sneeze to a distance of 2 metres (about 6 feet), and infective droplets can be spread over this distance. The bacteria enter through the nares or through open wounds. The infection then spreads to neighbouring tissues, or to distant sites haematogenously. Infection travels from the nares to the middle ear via the eustachian tube. Rarely, the bacterium can be transferred at mating, or parturition.

Diagnosis: Nasal swabs of rabbits with 'snuffles' may reveal *Pasteurella.* However, in the case of infection at distant sites nasal culture may be negative. *Pasteurella* can also be cultured from the respiratory tracts of healthy rabbits. Serology (ELISA) can be done. A high antibody titre may imply a chronic carrier state rather than the elimination of infection, so the results of serology and culture must be assessed alongside the clinical signs.

Radiography may be necessary to determine the extent of respiratory disease. In cases of otitis the tympanic bullae which are normally thin walled and hollow develop an increased soft tissue opacity and thicker walls.

Post-mortem examination of cases of sudden death associated with *Pasteurella* have petechiation and microscopic abscesses throughout the viscera.

Treatment: It is not usually possible to cure infected rabbits, but it may be possible to stabilise them with antibiotics so that they can live with chronic disease. Antibiotics are frequently needed for long periods, often as long as 6 months to 1 year. A culture should be done where possible. Pasteurella is generally sensitive to enrofloxacin, chloramphenicol, gentamycin, tetracycline and trimethoprim-sulpha drugs. Enrofloxacin is the drug of choice for long term medication, and can be given initially at a dose of 5-15 mg/kg twice daily, and then given orally at 10 mg/kg daily, or via the drinking water at a concentration of 50-100 mg/litre. Severe cases can be given procaine penicillin at a dose of 20 000- 60 000 iu daily for 5 days by intramuscular injection.

Eye preparations containing antibiotic can also be used. Gentamicin can be used as ear or nasal drops as indicated. Fluids are important, and can be given subcutaneously at a rate of 100 ml/kg per day in debilitated rabbits. The fluids will help hydrate the mucus produced by the nasal epithelium and facilitate its excretion. A good diet is very important, and the addition of vitamin C may improve recovery. The rabbit should be kept at a constant environmental temperature of 168C (about 608F), with a relative humidity of between 50 and 70%.

Prevention: Good husbandry is very important. A good diet of hay and fresh greens will protect against infection by improving the rabbit's resistance. In a group where *Pasteurella* is endemic, early weaning at 4-5 weeks will remove the young before they become infected. However, weaning at this time can be stressful. If weaning later, antibiotics can be given over the time of weaning, and tetracycline can be given in the drinking water at a dilution of 250 mg/litre.

Rabbits can sneeze over a distance of 2 metres (about 6 feet), and where possible hutches should be spaced apart to prevent spread by infective droplets.

Vaccination with a sheep vaccine may help control the disease, but will not cure it. Vaccination is good at stimulating a non-specific immunity and may give the rabbit a better resistance against other infections.

2. Bordetella bronchiseptica

Many rabbits carry *Bordetella bronchiseptica* in their upper respiratory tract. It is not associated with disease, but its presence may increase the rabbit's susceptibility to *Pasteurella* infection. It can be transmitted to guinea pigs and it is pathogenic in this species; *B. bronchiseptica* infection should be considered if rabbits are housed with guinea pigs.

3. Staphylococcus aureus

This bacterium is carried in the upper respiratory tract of healthy rabbits, and it is also carried by humans. It may cause disease depending upon the virulence of the strain and host resistance. It is the commonest cause of conjunctivitis, and may also be associated with the development of a nasal discharge and pneumonia. The pneumonia is associated with the formation of multiple small abscesses in the lungs, in comparison with *Pasteurella* pneumonia which causes a purulent pleuropneumonia. When possible, treatment should be based on a bacterial culture and sensitivity test. If this is not possible the best antibiotics are enrofloxacin or a trimethoprim-sulpha combination. Antibiotics will reduce the clinical signs but may not eliminate the organism.

Eye ointment containing fusidic acid is active against *Staphylococcus* and is useful in cases of conjunctivitis.

4. Moraxella catarrhalis (Formerly neisseria)

This bacterium is commonly carried in the upper respiratory tract of rabbits. It is not considered a primary pathogen, but may cause infection as an opportunist if the respiratory mucosa is already diseased. No treatment is necessary if it is isolated from the nares of a healthy rabbit.

5. Pseudomonas

Pseudomonas is not a normal inhabitant of the respiratory tract, and if it is isolated it should be considered pathogenic. It is an opportunist bacterium which will colonize weakened respiratory mucosa, and predisposing environmental factors must be investigated.

B. Viral Diseases

Viral causes of respiratory disease are not well documented, and thought not to be a common problem. The myxoma virus causing myxomatosis is accompanied by respiratory symptoms only if there is secondary bacterial infection in chronic cases. Mild cases of haemorrhagic viral disease (HVD) which recover may go on to develop secondary respiratory disease. HVD causes multiple lung haemorrhages, and death in its more acute form.

7.5. The Digestive System

A. The Mouth

1. Oral papillomatosis

This is a rare condition caused by a papilloma virus. White lesions occur on the underside of the tongue and may become ulcerated. The condition is usually self-limiting and the lesions regress after several weeks.

2. Oral ulceration

Ulceration in the mouth is common, and the first presenting signs are salivation and anorexia. The ulcerations are caused by the overgrowth of the cheek teeth. The lower molars cause lingual ulceration, whilst spurs from the upper molars generally grow laterally and cause buccal ulceration. Treatment is aimed at correcting the underlying dental problem.

B. The Stomach and Small Intestine

1. Gastric stasis

If gastric motility is reduced, ingesta remains in the stomach for longer. If hair is mixed up with the ingesta, a solid impacted mass forms as the stomach contents dehydrate. Such a 'hairball' is not a primary condition, but secondary to reduced gut motility. Reduced gut motility is usually the result of feeding a low fibre diet. In these cases the stomach and caecum empty very slowly. The rabbit produces less and less pellets, and will stop eating. A mass of impacted ingesta is generally palpable in the stomach and caecum. Radiographic signs include a classic 'halo' of gas around a ball of impacted foodstuff in the stomach.

Treatment

- Fluids can be given by subcutaneous injection, and orally (with an electrolyte solution).

- Pineapple juice will provide fluids, and the enzyme (bromelain) will help break down the mucus that binds the hair and ingesta together in the impaction. Up to 10 ml can be given three times a day.
- A cat laxative can be given 30 minutes after the pineapple juice. The dose is 3ml twice daily. Alternatively liquid paraffin can be given at a dose of 5ml two or three times daily.
- Metoclopramide given by subcutaneous injection will stimulate gastric emptying and gastric motility.
- Cisapride 0.5-1 mg/kg can be given orally every 8-24 hours.
- Vitamin B injection, as ingestion of caecotrophes is unlikely to be taking place.
- Leafy greens should be offered, especially those of laxative plants, e.g. dandelions. Also offer good quality hay.
- Advanced cases may require surgery to relieve the impaction, but only if symptomatic treatment and supportive care have made no difference. Depending upon the condition of the rabbit it may take 2-3 days for the medical treatment to stimulate gastrointestinal motility, and only when this fails should surgery be considered.
- Analgesia.
- Antibiosis with a trimethoprim-sulpha combination at a dose of 30 mg/kg twice daily will help prevent bacterial overgrowth.

Prevention: A high fibre diet should be fed of good hay and a reduced pelleted ration. The pellets should contain a minimum of 18%fibre and less than 16% protein. A variety of vegetables should be gradually introduced into the diet, and then included on a daily basis. Starch and fat should be minimised (no legumes, grains, etc.) Rabbits should be allowed to exercise daily.

2. Hairballs (Trichobezoars)

Hairballs are no longer thought to be a primary condition, but rather to occur secondary to gut stasis. There are, however, predisposing factors that make their formation most likely.

- A low fibre diet (high fibre diets stimulate the rapid passage of food through the digestive tract, taking the hair with it).
- Low exercise.

- Long haired breeds, or any breed during an excessive moult. (Rabbits under 7 months of age are unlikely to suffer hairballs as it is only at this time that they develop a full hair coat.). Mutual grooming of a companion that is moulting.
- House rabbits can ingest other fibres such as carpet which will cause a similar impaction.

Treatment: Same as for gastric stasis.

Prevention: A high fibre diet and increased exercise. Exercise stimulates peristalsis (gut motility) and improves the passage of food through the gut. The grooming of long-haired breeds and those in heavy moult will help.

3. Gastric intestinal obstruction

Obstructions may occur at the pylorus of the stomach, or in the duodenum. The cause of the obstruction may be matted hair (matted before ingestion), plastic foreign bodies, or ingested carpet. Ingestion of clay cat litter can also cause obstruction.

Clinical signs: Sudden onset gastric pain and bloating. On radiography, gas accumulation will be evident cranial to the obstruction.

Treatment: An analgesic must be given such as buprenorphine at a dose of 0.01-0.05 mg/kg by intramuscular injection. A dose of rapid acting corticosteroid is indicated to counteract shock. Fluids should be given, preferably intravenously. The obstruction must be removed surgically.

4. Intussusception

This is uncommon, but may occur in young rabbits in association with enteritis, or in older rabbits associated with a foreign body.

Clinical signs: Anorexia, and a cessation of droppings. The intussusceptions is generally palpable on abdominal examination, and radiographs will demonstrate the build up of gas cranial to it.

Treatment: Surgical correction. Warmed fluids can be given intraperitoneally as the abdomen is being closed.

C. The Caecum

1. Excess caecotrophs ('sticky bottom syndrome')

Some diets, particularly those that are high in protein, low in fibre and high in carbohydrate or sugar, will cause the rabbit to produce more caecal faeces than it needs, and hence develop a 'sticky bottom'. If the rabbit is overfed, particularly if it is on a high protein diet, it may lose the urge to practice coprophagy. Large rabbits kept in small hutches may not have space to stand on their hind legs and reach their anus. Caecal faeces may also build up if the rabbit is unable to perform coprophagy, if for example it is overweight, has an excessive dewlap, has incisor or molar malocclusion (which will make coprophagy uncomfortable) or has arthritis and spinal pain.

Dietary reform means changing the rabbit over to a high fibre, low protein and low carbohydrate diet, which actually provides a more natural diet, and one closer to that of their relatives, the wild rabbit. The rabbit's digestive system works at its best when it has to break down plant fibres.

(1) There should be unlimited access to good quality grass hay at all times, as the major fibre provider.

(2) Dry or pelleted food (which is high in protein and carbohydrate) should be restricted.

(3) Processed 'treats' (full of carbohydrate and sugar) should be cut out and replaced with fruit treats.

(4) Vegetables and fruit can be slowly introduced from the following list, one at a time. If any item upsets the rabbit it will show diarrhoea 24-48 hours after its introduction, and that fruit or vegetable should be removed from the diet. Once the new regime is fully introduced the rabbit should receive at least 12 cup of vegetables per kilogram of body weight daily. Vegetables: broccoli, brussels sprouts and sprout tops, cabbage and cabbage types, spring greens etc., carrot and carrot tops, celery, clover, dandelions (leaves and flowers) - use sparingly, kale, mint, parsley, radish tops, spinach, watercress. Fruit: small amounts of apple, peach, pear, melon, pineapple, plums, strawberries and tomatoes can be added or used as 'treats'.

(5) Astringent plants such as blackberry or raspberry leaves can be introduced immediately; they are a good source of fibre and will not cause diarrhoea.

Hay alone should be fed first, provided the rabbit is used to eating hay, otherwise it must learn to eat hay before everything else is withdrawn. After the faeces have been normal for a week, the greens can be introduced slowly. It may take a week, or up to several months on hay only before the faeces are normal.

2. Caecal impaction/tympany

If the passage of ingesta is slow, and if the diet is a low fibre one, this will result in the build-up of food material in the caecum. Prolonged fermentation of this material will lead to tympany, and if there are excess sugars or starch present this will alter the pH of the caecum and allow a build up of the bacterium *Clostridium spiriformes* which can lead to a fatal enterotoxaemia. Caecal impaction can be easily palpated per abdomen.

Treatment

- Gut motility must be encouraged as for gastric stasis, with and cisapride.
- A probiotic provides *Lactobacillus*. Although *Lactobacillus* and *Acidophilus* are not significant microorganisms in the rabbit caecum, in situations where a build-up of harmful bacteria is likely, their presence may help overpower the *Clostridium spp*. The Lactobacillus is encapsulated, and able to pass through the stomach's acidity into the caecum.
- Transfaunation: This describes the process of giving a faecal cocktail from a healthy rabbit to a sick one, in order to replace the natural micro-organisms, supply B vitamins, and stimulate the appetite.
- A buster (Elizabethan) collar is put around the healthy rabbit to prevent coprophagy, and some caecal pellets are harvested overnight. One faecal pellet can be mixed with 5 ml warm water, and administered orally by syringe or stomach tube.

3. Enteritis

Enteritis (diarrhoea) can occur for many reasons; a dietary change, bacterial, protozoal, viral, following antibiotic use, or at times of stress. Specific syndromes are detailed in this section. Rabbits with diarrhoea often have a higher stomach pH (3-7), enabling bacteria to pass to the hind-gut.

Basic treatment: Whatever the cause of enteritis, the principles of treatment remain the same.

- Nursing. The patient should be kept in warm, clean, stress-free surroundings.

- Fluid replacement. Aim to provide between 50 and 100 ml/kg per day in divided doses. Fluids (lactated Ringer's solution (Hartmann's solution)) can be given by intravenous or subcutaneous injection. Oral fluid replacement can be given by syringe or stomach tube.
- Stimulate normal gut motility. Metoclopramide can be given at a dose of 0.5 mg/kg up to four times a day; this stimulates gastric emptying, and should lead to an improvement of caecal motility.
- Antibiotics.
- Analgesia.
- Probiotics.
- Transfaunation. The permanent way to repopulate the caecum.
- Diet. Astringent plants such as shepherd's purse, blackberry and raspberry leaves can be given. The fibre content of the diet should be increased: leave off dry food and feed good quality hay. Herbal powders made from astringent plants can be used.
- Kaolin given as a kaolin or kaolin-pectin mixture orally may help bind up the liquid faeces.

4. Neonatal (Colibacillosis)

Affects rabbits of 1-14 days of age. Diarrhoea in neonates is commonly caused by an enteropathogenic *Escherichia coli*. Rabbit strains of E. coli do not produce enterotoxins as they do in other species.

Clinical signs: A yellow, watery diarrhoea which stains the perineum and abdomen. The morbidity and mortality within a litter may reach 100%.

Pathology: At post-mortem there is milk in the stomach, whilst the rest of the gastrointestinal tract is filled with watery contents. The finding of *E. coli* is significant in rabbits under 2 weeks of age, as it is not a normal component of the gastrointestinal flora at this age.

Treatment: Early cases may respond to antibiotic therapy (either trimethoprim-sulpha combinations or neomycin) fluids and a probiotic. Subsequent litters from the same doe may have passive immunity. As poor husbandry and sanitation are predisposing factors for colibacillosis, these must be investigated and improved as necessary.

5. Weaning enteritis

Affects rabbits of 4-8 weeks of age. At weaning the body undergoes various physiological changes. In suckling rabbits the pH of the stomach is between 3 and 6.5, whereas in the adult the pH is 1-2. For the first 21 days of its life the kit is protected by an antimicrobial fatty acid (the stomach oil) which is produced by the gastric lining in the presence of doe's milk and which controls its gastrointestinal microflora. After 21 days this protective factor wanes and the hind gut becomes populated with bacteria. A high stomach pH means that harmful bacteria may be able to pass through the stomach into the caecum and small intestines, and cause enteritis. In association with this, if the weanling consumes large amounts of high energy food, this will provide fuel for the multiplication of harmful bacteria (*Clostridium spiriformes*) in the caecum. At this time the rabbit is more dependent than ever on a healthy population of micro-organisms in the caecum.

Clinical signs: Rabbits of 4-8 weeks of age are commonly affected. There is profuse diarrhoea, dehydration and a rough hair coat. Mortality may be 100%.

Pathology: Lesions are commonly seen in the caecum. There is haemorrhagic inflammation of the caecum which is filled with watery or mucoid blood-tinged fluid.

Treatment: Fluid replacement with subcutaneous fluids (lactated Ringer's solution (Hartmann's solution)); oral fluids with an electrolyte solution. Fluid replacement should be maintained at a rate of 50-100 ml/kg per day in divided doses. A probiotic can be given to populate the gut with *Lactobacillus* and *Enterococcus*, or the gut can be repopulated with caecal pellets from a healthy rabbit.

A broad-spectrum antibiotic can be given, e.g. sulphonamides or neomycin. During an outbreak the pellet ration should be decreased, and unlimited good quality hay offered. Water should be available at all times.

Prevention: During the period of weaning plenty of fibre (hay) should be available. The concentrate ration should be introduced slowly so that there is not a flood of energy and nutrients into the caecum which may destroy the balance of micro-organisms. A probiotic can be dissolved in the water during this stressful time.

Stress should be minimised. In the case of intensive breeding this may mean the provision of larger hutches, with a separate nest area to avoid overcrowding. It may be advisable to leave the young with the doe longer before weaning, giving their immune system longer to develop. When weaning the doe should be removed from the young, rather than the young being taken away. This avoids the double stress on the young of a new environment and no mother. Concurrent infection with coccidia should be avoided.

6. Mucoid enteritis

Affects rabbits of 7-14 weeks of age. The exact cause of mucoid enteropathy is unknown, but its aetiology is similar to that of weaning enteritis. It is caused by changes in caecal pH that are associated with disruption of the normal caecal flora. It is a disease of young rabbits, just beyond weaning age (7-14 weeks) which is the age when the caecal microflora are becoming established, and are at their most vulnerable. Mortality is common. In up to 60% of cases there may be an accompanying pneumonia.

In older rabbits an enteritis associated with mucus production is more likely to be part of the 'enteritis complex' rather than classic 'mucoid enteritis' and mortality is lower. Mucoid enteritis is rarely seen in rabbitries that feed a high fibre diet and avoid excesses of grains, proteins and fats.

Clinical signs: Anorexia, lethargy, weight loss, diarrhoea, polydipsia, caecal impaction and an excessive production of mucus by the caecum.

Treatment

As for weaning enteritis.

- Fluid replacement, either via subcutaneous injection, or orally with an electrolyte solution.
- A high fibre diet is offered, good grass hay or alfalfa. Vegetable baby food can be given orally by syringe feeding.
- Metoclopramide can be given at a dose of 0.5 mg/kg up to four times a day: this stimulates gastric emptying, and should lead to an improvement of caecal motility.
- Cisapride is another motility drug and can be given at a dose of 0.5-1 mg/kg every 8-24 hours.
- The caecal flora should either be re-established permanently by transfaunation, or temporarily protected with Lactobacillus in a probiotic.

Prevention: The provision of a high fibre diet, and a gradual introduction to the pelleted ration at weaning. A probiotic can be dissolved in the drinking water at this vulnerable time. The use of tetracyclines at a concentration of 250 mg/litre in the drinking water for the first 10 days after weaning has been reported.

7. Enterotoxaemia

Affects rabbits of any age. This describes the harmful consequence of the normal balance of microorganisms in the caecum becoming disturbed. The balance tips in favour of harmful micro-organisms (*Clostridium spiriformes*). These bacteria multiply rapidly, producing endotoxins (iotatoxins) which cause rapid dehydration, collapse and death. Factors which lead to Clostridium proliferation are the following:

- An alteration in the caecal pH: normal pH is 5.9-6.8. If excess fermentable carbohydrates are fed, the caecum becomes more acidic and favours clostridial multiplication.
- A falling fibre content of the diet leads to a rise in populations of *E. coli* and *Clostridium*. Low fibre diets lead to a fall in volatile fatty acid production, and a rise in pH which unbalances the microflora.
- Oral administration of antibiotics (and to a lesser extent injectable antibiotics) will alter the normal balance of microflora.
- Excess dietary protein will lead to ammonia production in the caecum which will destabilise the caecal flora.
- Stress of any kind, e.g. weaning, a change of environment, hypothermia, will cause the release of adrenaline. This reduces gut motility and caecotroph production, which predisposes to Clostridium proliferation.
- Poor sanitation

Clinical signs: Anorexia and depression. Initially there is diarrhoea which is brown, watery and may contain blood or mucus. The rabbit soon becomes hypothermic and moribund. Death occurs within 24-48 hours. Some cases may become anorexic, bloat and die before diarrhoea develops. The abdomen sounds full of fluid if the rabbit is moved.

Diagnosis: Diagnosis is based on the history and clinical signs. At postmortem there may be haemorrhages on the serosal surface of the caecum, and haemorrhages and mucus present on the mucosal surface of the caecum and proximal colon. Gut stasis leads to the accumulation of gas throughout the digestive tract.

Treatment: The prognosis of these cases is guarded. Aggressive fluid therapy is essential. Metronidazole at a dose rate of 20 mg/kg orally twice daily may be of benefit. Repopulation of the intestinal flora can be achieved temporarily with a probiotic, or more permanently by transfaunation. Vitamin C (ascorbic acid) may inhibit toxin production, and can be given at a dose of 50-100 mg/kg daily.

Prevention: The feeding of a high fibre hay-based diet is extremely important, particularly at any stressful or vulnerable time. The concentrated food should be rationed to prevent protein and carbohydrate overload. Vitamin C and a probiotic can be given at times of stress. The use of a sheep vaccine has been reported in an outbreak of Clostridium perfringens.

8. Colibacillosis in older rabbits

Affects rabbits of any age post-weaning. This disease can occur in rabbitries where there is poor husbandry and sanitation. It can also be a problem associated with pet shops where young rabbits from several sources may be mixed together.

Clinical signs: These depend upon the strain of E. coli involved, and range from a mild diarrhoea and weight loss, to a severe form of diarrhea associated with up to 50% mortality. Affected rabbits may develop an intussusception or rectal prolapse. Recovered animals may remain undersized.

Diagnosis: Isolation of E. coli from stool or tissue samples may suggest colibacillosis; however, non-pathogenic E. coli can be isolated from any rabbit, and in greater numbers from a rabbit with caecal dysbiosis. At postmortem classic 'paintbrush' haemorrhages may be seen on the caecal wall. The colon will also be affected.

Treatment: Antibiotics (either trimethoprim-sulpha combinations or neomycin) fluids and a probiotic should be used. The fibre content of the diet should be increased.

9. Enterocolitis syndrome

This is a new disease condition first identified in France in 1997, and possibly seen in Belgium, Spain and Portugal. The syndrome was first seen in young rabbits of 5-7 weeks of age, but has also been reported in weanlings and lactating does. The disease appears to be highly contagious, and although the exact cause has not been identified it is thought it may be associated with a viral infection.

Clinical signs: Anorexia, a watery diarrhoea which is not profuse, reduced gastric motility progressing to ileus and death after 2-3 days. At postmortem there is abundant liquid in the small intestine, and gas and mucus in the colon.

10. Tyzzer's disease

Affects rabbits of any age. Not common in pet rabbits. This is caused by Clostridium piliforme (formerly *Bacillus piliformis*). Other rodents are affected, and wild vermin may act as carriers of the disease. Any form of stress increases the rabbit's susceptibility to infection.

Clinical signs: Symptoms may be acute in the young rabbit. Signs include watery diarrhoea, listlessness, dehydration, anorexia and death. Older rabbits may develop chronic weight loss.

Diagnosis: At post-mortem there may be the classical focal necrosis of the liver, however, this is a less common finding than in other rodents. Silver staining of these lesions will identify the bacterium. The myocardium and gastrointestinal tract may have similar areas of necrosis. There may be oedema of the intestinal walls, especially in the caecum.

Treatment: There is a poor response to antibiotic therapy. Tetracyclines are the antibiotic of choice, and these can be given in the drinking water during an outbreak at a concentration of 125 mg/litre for up to a month.

Prevention: This is most important: stress should be minimised, and the rabbits should be fed on a high fibre (hay) diet. Wild rodents should be prevented from entering the environment, or contaminating the food. The bacterial spores will be killed by a 0.3% sodium hypochlorite solution.

11. Salmonellosis

This is uncommon, but can occur in association with contaminated food or water. It is generally caused by *Salmonella typhimurium*.

Clinical signs: Septicaemia, depression, pyrexia and death, often accompanied by diarrhoea. Does may develop metritis, and pregnant does may abort their young.

Diagnosis: Post-mortem findings reflect the septicaemia, diffuse petechial haemorrhages, and vascular congestion of organs. Necrotic foci are found on the liver and spleen. The lymph nodes may be oedematous.

Treatment: None. The disease is a zoonosis. Because of the public health risk affected animals should be destroyed, bedding and hay should be burnt, and all utensils thoroughly disinfected.

12. Enteritis complex

Affects older rabbits. Older rabbits can also show symptoms similar to the mucoid enteritis of young rabbits. The clinical signs may range from an intermittent diarrhoea, with stools covered in mucus, to a more severe diarrhoea which precipitates enterotoxaemia. Excess mucus is formed by both the caecal and colonic lumen. Any changes which upset the balance of the caecal flora can be responsible for triggering this condition. Thus sudden dietary changes, stress, antibiosis or gut stasis can cause enteritis.

Treatment: This depends on the severity of the condition. Mild cases can be corrected by dietary changes such as increasing the fibre and decreasing the carbohydrate and protein content of the diet. More severe cases may need the same treatment as for mucoid enteropathy.

13. Antibiotic-induced diarrhoea

Oral administration of antibiotics, and to a lesser extent injectable antibiotics can alter the normal flora of the gastrointestinal tract, and allow *Clostridium spp.* to proliferate, causing enterotoxaemia. The most toxic drugs are those with a narrow spectrum of action against Gram-positive organisms such as penicillin, erythromycin, lincomycin and clindamycin. The negative effects of any antibiotic should be minimised by the concurrent use of a probiotic, and by ensuring that the rabbit is on a hayrich high fibre diet.

14. Chronic diarrhoea

Some rabbits may present with diarrhoea of several months' duration. This must be differentiated from the build up of excess caecotrophs. There may be no hard pellets produced, and the liquid faeces will not have the strong smell that caecotrophs have.

Treatment: The diet should be changed to hay and astringent plants alone. All pelleted food and treats should be removed. Probiotics are particularly useful.

15. Dietary enteritis

Any sudden change in diet can cause enteritis. Overfeeding can also cause enteritis. The carbohydrate overload theory' suggests that if there is too much grain starch in the diet the digestive capacity of the stomach and small intestine is exceeded, and undigested starch reaches the caecum, where it can alter the caecal pH, upset the microflora, and cause enteritis.

Diets and treats made with starch and refined sugars will have the same effect. If large balls of concentrated foodstuffs enter the stomach they may not be effectively sterilised, allowing potentially harmful bacteria to enter the hind-gut.

Treatment: Provision of fibre is very important, and up to 20% of the diet should consist of coarse fibre. The rabbit should be provided with good hay only, plenty of fluids (given orally, or via subcutaneous injection), and the balance of microflora re-established, either with a probiotic, or by transfaunation. Astringent plants can be introduced to the diet before re-introducing the dry food.

D. Viral Diseases

1. Rotavirus

Infection with rotavirus alone may be subclinical, and it may be endemic in some situations. However, in association with stress and secondary infection it can produce severe clinical signs of anorexia, diarrhoea, dehydration and death. Suckling and weanling rabbits between 4 and 12 weeks of age are most vulnerable.

Pathology: There is villous atrophy of the jejunum and ileum. The intestines are distended and congested, and there are petechial haemorrhages in the small and large intestines.

Diagnosis: This is by virus isolation.

Prevention: This is difficult as the virus may be endemic. Stress of all forms should be reduced, and the diet should be improved to include more fibre in the form of hay.

2. Coronavirus

A coronavirus has been identified that causes disease in rabbits between 3 and 10 weeks of age. In this age group the clinical signs are severe anorexia, diarrhoea, dehydration and death, whilst adult rabbits can carry the virus subclinically.

Pathology: This is similar to that produced by a rotavirus.

Diagnosis: This is via virus isolation.

3. Calicivirus

Viral haemorrhagic disease is caused by a calicivirus.

E. Endoparasites

1. Stomach worm

This is found in wild rabbits, and occasionally in pet rabbits. The worms (*Graphidium strigosum*) are red, 1-2 cm long, and found on the stomach lining. In large numbers they may cause weight loss and death.

Treatment: A roundworm product such as fenbendazole can be used at a dose of 10-20 mg/kg orally. Ivermectin given by subcutaneous injection at a dose of 0.4 mg/kg is also effective.

2. Pinworm

These worms (*Passalurus ambiguous*) are thought to be non-pathogenic. They are very small (0.5 cm long) and found in the rectum and stuck by the anus. They have a direct life-cycle; the adults live in the caecum and colon, and eggs are passed in the faeces.

Treatment: Fenbendazole can be used at a dose of 10-20 mg/kg orally, and repeated in 14 days. Treatment is not strictly necessary as the worms are non-pathogenic, but owners may prefer not to see them. Ivermectin is also effective against pinworms at a dose of 0.4 mg/kg.

3. Tapeworm

These are considered to be very rare. The rabbit is the definitive host of *Mosgovoyia pectinata,* the adult stage of which is found in the small intestine, and segments of which are passed in the faeces. The eggs from the segments are eaten by free living mites in which the worm embryo develops. Mites containing the infective larval stage are then ingested by rabbits as they graze. The rabbit is also an intermediate host for *Taenia pisiformis* and *Taenia serialis.* The dog is the definitive host for these tapeworms, and the rabbit can become infected if it grazes on vegetation soiled by dog faeces. The intermediate stages can be found in infected rabbits: cysticerci in the liver and mesentery, and coenuri in the muscle and connective tissue.

Treatment: Praziquantel can be given by subcutaneous injection at a dose of 5-10 mg/kg. This can be repeated after 10 days.

4. Coccidiosis

Coccidia are the commonest parasites of the rabbit. Twelve forms of this protozoal parasite are known to infect the rabbit, all from the genus *Eimeria*. One species, *Eimeria stiedai* parasitises the liver, whilst the rest colonise the gastrointestinal tract. Not all species are pathogenic. The most common intestinal species are *Eimeria perforans, Eimeria magna* and *Eimeria media.*

5. Hepatic coccidiosis

This is caused by *Eimeria stiedae*. The parasite colonises the biliary epithelium, and the oocytes produced pass down the bile duct and into the faeces. Sporulation requires 2 days or more outside the host. Infection is spread via the faeco-oral route, by ingestion of sporulated oocysts. Caecotrophs, however, are not part of this cycle as caecotrophy does not allow time for sporulation and the caecotrophs never contain oocysts.

Clinical signs: In mild infections young rabbits may just show general unthriftiness, or the disease may be fatal. Heavily infected rabbits show symptoms associated with bile duct blockage, an enlarged and failing liver, ascites, jaundice, diarrhoea, anorexia and death.

Diagnosis: The presence of coccidial oocysts does not always confirm the diagnosis, as many rabbits are subclinically infected with coccidia. However, the isolation of large numbers of coccidia, and the presence of clinical signs will suggest coccidiosis. At post-mortem the liver is enlarged and covered with yellow-white nodules. When these nodules are cut they exude a yellow-green fluid. The same fluid may be found in the gall bladder and bile duct. The number of foci is related to the number of infected oocysts ingested. Histopathological examination of the liver will confirm the diagnosis.

Treatment: In large rabbitries, pellets may be fed that contain a coccidiostat. However, strains of coccidia can become resistant to these drugs; they also appear less effective against hepatic than intestinal coccidia. Sulpha drugs are the most effective. Sulphadimidine can be given in the drinking water at a concentration of 0.2%. To achieve this concentration 1ml of Sulphamezathine 33% can be diluted in 150 ml water (i.e. 6 ml/litre). This can be given as three 3 day courses, with two 2 day intervals in between. Sulphadimethoxine can be given orally at a dose of 25 mg/kg daily for 10-14 days. Alternatively a trimethoprim-sulpha combination can be given by mouth at a dose of 30 mg/kg twice daily. The combined amount of trimethoprim and sulpha drug is used when working out the dose rate. The treatment should be continued for 10-14 days. Amprolium 9.6% in the drinking water is also effective. Attention to hutch sanitation is extremely important because the oocysts thrive in damp conditions. Hutches should be cleaned and disinfected regularly, as the oocysts can remain infectious in the environment for several months.

6. Intestinal coccidiosis

The life cycle of these *Eimeria* species is similar to *Eimeria stiedae,* except that the ingested oocysts colonise the intestines. The site and pathogenicity depend upon the species. *Eimeria magna* is highly pathogenic of the small intestine, *Eimeria perforans* is of low pathogenicity of the small intestine, and *Eimeria media* is of moderate pathogenicity of both large and small intestines. Mixed infections are common.

Clinical signs: These depend upon the species, and on the quantity of infective oocysts that are ingested. Mild infections are unapparent, and a previous exposure leads to the development of immunity. Clinical signs are most common in young rabbits, and include diarrhoea, weight loss, anorexia and, in severe cases, dehydration and death due to secondary bacteraemia. The diarrhoea frequently contains blood and mucus. If the diarrhoea is severe it may lead to the development of intussusception.

Diagnosis: Oocysts will be found on faecal examination; however, oocysts can also be found in healthy rabbits, so results must be interpreted with care. At post-mortem there are ulcerative lesions of the intestinal lumen. Definitive diagnosis is by histopathological examination of the intestines.

Treatment: Same as hepatic coccidiosis.

Prevention of coccidiosis: Elimination of *coccidia* is difficult, as many rabbits may develop an immunity following previous exposure and become carriers, particularly in the case of hepatic coccidiosis. Attention to hutch sanitation is most important. The oocysts need at least 2 days in the environment to sporulate and become infective, and they thrive in damp conditions; therefore regular cleaning of the hutch will remove these oocysts. As oocysts can survive for many months in the environment, deep-litter situations are not recommended, and if the rabbit is kept in an ark on the grass it should be moved regularly. Stress of any sort such as overcrowding or poor nutrition make rabbits more susceptible to clinical disease, and any stress factors must be reduced.

7. Yeasts

The yeast Saccharomyces is a normal inhabitant of the caecum. The numbers of this yeast vary with the composition of the diet; as the fibre content of the diet increases the number of Saccharomyces falls.

F. Anorexia

This may be associated with pain of any sort or stress. A full clinical examination is essential to determine the cause of the problem, and the following treatment protocol can be instigated whilst investigations are underway. The aim of treatment is to provide enough calories to prevent hepatic lipidosis, to rehydrate the gastrointestinal contents, and to provide indigestible fibre to help promote peristalsis.

Treatment: Keep the rabbit in an environmental temperature of less than 24°C (75°F).

- Fluid replacement. Lactated Ringer's solution (Hartmann's solution) can be given by subcutaneous injection at 100 ml/ kg per day.
- Analgesics.
- Offer leafy greens, dandelions, parsley, carrot tops, or kale immediately. Fresh hay should also be available.
- Syringe feeding. A mixture of pineapple juice, ground pellets and strained vegetable baby food in a 1:1:1 ratio can be given. 10-15 ml/kg of this mixture can be fed two or three times a day. Canned pumpkin and alfalfa powder, if available, can be added to the mixture. Force feeding is very important to correct or prevent hepatic lipidosis which occurs very quickly after a rabbit stops eating. If necessary the rabbit should be fed through a nasogastric tube.
- Metoclopramide can be given at a dose of 0.5 mg/kg up to four times daily to stimulate normal gut motility.
- Vitamin C 50-100 mg/kg twice daily, together with vitamin B (1-2 mg/ kg thiamine) if caecotrophy is impaired; in the latter case Multivet 4BC injection (C-Vet) can be given at a dose of 0.3 ml/kg.
- Antibiotic therapy may be given if indicated. Enrofloxacin or a trimethoprim -sulpha combination are safest. The latter can help protect against caecal dysbiosis.
- A probiotic will help prevent caecal dysbiosis.

G. Lead Poisoning

This is included here as it should be in the differential diagnosis for anorexia, particularly since more rabbits are kept free range indoors with potential access to lead in old paint, curtain weights, the foil around wine bottles, etc.

Clinical signs: These may be vague and include anorexia, lethargy and neurological changes. Anorexia and weight loss are the most common symptoms.

Diagnosis: An abdominal radiograph may show ingested metal in the abdomen. A blood sample for lead levels should be taken, and haematology may reveal basophilic stippling of the red blood cells.

Treatment: Rabbits with a blood lead level greater than 10 *m*g/dL should be treated with injections of calcium versenate (Ca-EDTA). The dose is 27.5 mg/kg by subcutaneous injection four times a day for 5 days. A second course may be required a week later.

H. Obesity

Clinical significance

- Hepatic lipidosis
- Pododermatitis
- Chronic soft stools

Obesity is a real problem in pet rabbits. Most rabbits are fed a diet rich in carbohydrates and protein, and given extra treats, whilst getting little or no exercise. The key to managing obesity is to make gradual changes. The diet should be based on hay and leafy green foods, with no or little dry food. If the rabbit is already used to eating hay, the dry food can be removed completely; otherwise it should be removed gradually whilst the rabbit becomes accustomed to the hay and greens. If food intake declines rapidly the rabbit may develop hepatic lipidosis. A gradual weight loss of no more than 1-2% of the body weight per week should be the recommended target.

I. The Diseases of Liver

1. Hepatic lipidosis

The rabbit is very susceptible to the development of hepatic lipidosis. Even a short period of anorexia will lead to ketosis, and subsequent hepatic lipidosis. Ketonuria can be detected by the use of urine test strips. Liver biochemical parameters will be elevated.

Treatment: Rehydration with glucose-saline or lactated Ringer's solution (Hartmann's solution). If the rabbit is anorexic, force feeding is essential to maintain a positive energy balance.

2. Hepatic insufficiency

Pine and cedar shavings can cause liver disease. The removal of these from the litter tray will usually allow for a reversal of the condition to occur. Liver biochemical parameters will be raised; however, in severe hepatic insufficiency the BUN will be decreased, and the total protein will also be lowered.

J. The Diseases of Spleen

1. Yersiniosis

An enlarged spleen may be caused by infection with Yersinia pseudotuberculosis (rodentosis).

Clinical signs are non-specific and include cachexia and weight loss. The spleen is palpable on clinical examination. The bacterium is carried by rodents and birds, and is introduced to the rabbitry by contamination of food and bedding, and from the feeding of unwashed plants. Prevention is therefore aimed at controlling rodents and washing fresh feedstuffs.

Diagnosis of yersiniosis is not possible in the live rabbit. Post-mortem examination of infected rabbits reveals the enlarged spleen with necrotic foci on its surface, and similar foci in the lymphoid tissue at the ileo-caecal junction. The bacteria can be isolated from these foci. If yersiniosis is suspected in the live rabbit, treatment can be attempted with enrofloxacin at a dose of 5 mg/kg twice daily for 3-4 weeks.

2. Salmonellosis

In acute salmonellosis the spleen may be enlarged tenfold. Pathology reveals necrotic foci on the spleen, liver and Peyer's patches. There may be an associated metritis.

3. Toxoplasmosis

Splenomegaly is also seen in acute toxoplasmosis. The cat is the definitive host for the protozoan Toxoplasma gondii, and infection occurs if the rabbit eats food that is contaminated with cat faeces carrying the oocyst stage. Chronic infections remain unapparent, but the acute form causes anorexia, lethargy, fever, tremors, muscle weakness and paralysis. To avoid infection occurring, cats must be kept away from the rabbit's environment.

K. The Diseases of Pancreas

Diabetes is not generally considered as a clinical disease in rabbits, although it has been reported in New Zealand White rabbits. Experimentally rabbits have been shown to survive following a pancreactomy, suggesting that the role of the pancreas is not as important as in other mammals. Detection of glucose in the urine is a normal finding, and a raised blood glucose does not necessarily imply a diagnosis of diabetes as it may occur with excitement, hyperthermia, shock and in early mucoid enteritis. Rabbits that have a persistently raised blood glucose and are polydipsic can be managed on a hay-based high fibre diet. If the rabbit is obese the dry food can be cut out of the daily ration.

L. Neoplasia of the Gastrointestinal Tract

Various neoplasms have been identified. They may be primary lesions, or secondaries (generally from uterine adenocarcinomas). Some neoplasms may be amenable to surgical resection if identified early. Bile duct carcinomas are the exception as metastasis is common.

- Stomach: adenocarcinoma, leiomyosarcoma
- Intestines: leiomyoma, leiomyosacoma
- Sacculus rotundus: papilloma
- Bile duct: adenoma, carcinoma
- Rectum: anorectal papilloma.

7.6. The Musculoskeletal System

1. Pododermatitis (Sore hocks)

Clinical signs: Hair loss, scaling, erythema and ulceration of the skin on the plantar aspect of the metatarsus. Very occasionally the front feet may be affected too. The development of this condition is associated with abrasive or soiled cage floors, in conjunction with obesity and in rabbits that stamp frequently. In house rabbits it may be associated with 'carpet-burn'. It is common in rabbits that only have a thin covering of fur on the foot pad.

The lesions become secondarily infected with *Pasteurella multocida* or *Staphylococcus aureus*, and may develop caseous abscesses. Occasionally the condition may progress to cause osteomyelitis and septicaemia.

Treatment: The lesions should be cleaned with an antiseptic such as povidone-iodine. If the rabbit will tolerate dressings, the feet may be dressed. Dressings should be changed every 5 days, and once the hock is healing the area can be protected with a lighter covering such as a small sock. A soft, clean surface (either deep dry bedding or towels) should be provided.

The injection of a long-acting steroid under the lesion may be of benefit. The lesions can take weeks to heal. A tendency towards sore hocks may be inherited, and affected animals should not be bred from. Owners should be advised not to trim the fur pads from the point of the hock, but rather to wash the hocks in warm water, as fur trimming, especially in Angoras can predispose towards sore hocks.

2. Fractures

Rabbits readily fracture their legs. Their habit of stamping with their hind legs, or kicking out when they struggle means that they are also susceptible to vertebral fractures, often in the region of L6-L7. Rabbits that are selective feeders, and may have a low calcium intake, are more susceptible to fractures.

Diagnosis of fractures is by radiography.

Fracture repair

Fractured legs can be managed in the same way as for cats and dogs.

External fixation: The cast material must be light enough so that the rabbit is not hindered by the dressing. A half-cast and a light Robert Jones bandage may provide adequate support. Problems that may occur are the chewing of the dressing by the patient, or the dressing may become soiled, causing skin infections and pressure sores. Closed fractures should have a single dose of antibiotics at the time of stabilisation. External fixators have also been used to repair fractures. These need to be left in place for at least 6 weeks. Antibiotics should be given postoperatively.

Internal fixation: Many fractures can be repaired with intramedullary pins. Great care must be taken to debride and lavage open fracture wounds, as post-traumatic osteomyelitis is a common complication of infected fracture sites. Fractures managed by internal fixation need post-operative antibiotics for 7-10 days.

Post-operative Care

Whatever method is used for fracture fixation the post-operative care should be the same. Analgesia should be given at the time of stabilisation, and continued post-operatively as necessary. Post-operative antibiosis is important for open fractures to prevent the development of post-traumatic osteomyelitis.

The patient should be confined in a clean stress-free environment. Force feeding may be necessary if the rabbit is anorexic. Pain is often manifest as anorexia, so these patients should continue to receive analgesia. The fracture should be re-radiographed every 3-4 weeks to monitor the healing process. Callus formation may take as little as 10 days, and healing of an uncomplicated fracture is generally complete by 6 weeks.

3. Post-traumatic osteomyelitis

This is a potential complication of an open fracture, if the fracture site is contaminated or the vascular supply is damaged. Open fractures must be debrided and lavaged before repair to try and prevent this condition. Postoperative antibiosis, preferably based on the results of a culture, are very important and should be given for 7-10 days, or longer if necessary.

Radiographic signs of post-traumatic osteomyelitis include osteolysis and periosteal reactions around the fracture site. An abscess may develop at the fracture site, often associated with Pseudomonas infection.

Treatment with further antibiotics and wound lavage is generally unsuccessful, and amputation of the limb may be the only solution.

4. Osteomyelitis

Bacterial infection of bone can occur in the jaw, commonly associated with tooth root abscesses, and in the long bones secondary to pododermatitis, or nail-bed infections. Radiographically there is osteolysis and secondary new bone production associated with the area of infection. There may be drainage tracts in the bone associated with severe osteomyelitis.

Osteomyelitis is very difficult to treat, and requires surgical debridement of the affected area and the use of long-term antibiotics. Local impregnation of the affected area with antibiotics can be attempted. Where a limb is affected amputation may be necessary.

5. Amputation

Rabbits cope extremely well with the loss of a limb, either a fore or hind leg. The procedure for amputation is similar to that for a cat or dog. The forelimb is best amputated through the proximal humerus, whilst preserving as much muscle bulk as possible. The hind leg can be amputated at the proximal femur, whilst preserving the bulk of the quadriceps muscle. In smaller rabbits which may be lightly muscled, the femur is best removed at the coxo-femoral joint. Similarly if the hind limb is being amputated to resolve a congenitally dislocated hip, it can be removed at the coxo-femoral joint. Care must be taken when cutting the bones, as they can shatter easily, and are better cut with a rotating saw than bone-cutters. The muscle edges are inverted and sewn with a continuous suture using chromic cat gut, and skin closure is done with a continuous apposition suture. A single dose of post-operative antibiotics is necessary, and analgesia should be given for the first 24-48 hours.

6. Spinal Cord Injury

Traumatic vertebral injury with spinal cord damage commonly occurs in pet rabbits. The onset is acute and it may occur even after a slight struggle. Radiographs are necessary to assess the injury. Spinal fractures occur most commonly at L6 and L7. The prognosis depends upon the degree of spinal cord damage. Spinal cord injury in the absence of a vertebral fracture results in oedema around the cord. The rabbit should be given a rapid acting corticosteroid, and diuretics, and be cage rested. If some motor function to the hind limbs and bladder are retained at the time of injury, the prognosis for recovery is good. Nursing a rabbit with paresis or paralysis is very important. Food and water must be accessible, and the rabbit may need to be supported in a sitting position whilst it eats. The main complication is urinary and faecal soiling. The rabbit may need to be bathed regularly using a mild solution of chlorhexidine. The perineal area can be protected from urine scald by the use of a resistant barrier cream such as zinc and castor oil cream, or kept dry with the application of baby talcum powder. If urinary incontinence is a problem the rabbit can wear disposable nappies, or be bedded on an absorbent material such as 'Vetbed' or sheepskin. Care must be taken to avoid myiasis ('fly strike') if there is perineal soiling. Rabbits with spinal cord paresis will be unable to practice coprophagy. They should therefore be given a vitamin supplement, and weekly vitamin B injections. The caecotrophs they produce can be collected, and be fed back to them.

Geriatric rabbits can develop paresis and paralysis as a result of vertebral degeneration in association with ageing, and vertebral spondylosis is common. These rabbits should be nursed as above. Anti-inflammatory treatment can be given on a daily basis, and some patients may respond to complementary medicine such as acupuncture.

7. Arthritis

Reduced mobility in the older rabbit may be associated with arthritis. Radiographs will show marked periosteal proliferation in cases of degenerative joint disease. There may be associated soft tissue calcification or ossification. Analgesia can be given to control the pain. Aspirin as a paediatric syrup can be given at a dose of 100 mg/kg twice daily. Meloxicam can be also be given in liquid form. Long term corticosteroid use is contraindicated as it may activate subclinical Pasteurella infection. Complementary therapies such as acupuncture may be of benefit. If mobility is limited, the rabbit will be more prone to urinary and faecal soiling, and can be bedded on a soft absorbent bedding such as 'Vetbed'. Food and water should be made easily available, and a litter tray, if used, should have a low or absent rim so that it is easily accessible. If the rabbit is overweight it should be encouraged to lose weight slowly, by reducing its concentrate ration, and introducing more hay and greens.

8. Osteoporosis

This is a metabolic bone disease which results from diets that are too low in calcium. Rabbits can become selective feeders and those that leave the pelleted portion of their dry mix (which contains the calcium and vitamins) are most at risk. Lack of exercise can predispose to osteoporosis. Radiographically there is a decreased skeletal radiodensity. Radiographs of the spine will show indistinct transverse and dorsal spinous processes, and the vertebral end plates will be less dense than the vertebral bodies. Rabbits with osteoporosis are more prone to vertebral injury, and also dental malocclusion, as the poor mineralisation of the periodontal bone allows tooth movement in the sockets, and therefore malocclusion. Rabbits should be made to eat all parts of a complete ration. Less should be given, and the bowl should not be topped up until it is finished. A vitamin and mineral supplement can be sprinkled on the food. Rabbits should be allowed a minimum of 4 hours exercise a day.

9. Neoplasia

Osteosarcomas are generally found in the jaw bones, where they should be differentiated from osteomyelitis by radiography. Osteosarcomas have also been recorded on other parts of the appendicular skeleton. Osteochondromas are found on the long bones of the skeleton. Neoplasia carries a very poor prognosis.

10. Myositis

A severe myositis is associated with sarcocystis. The causative agent is *Sarcocystis cuniculi,* which forms cysts in the skeletal muscle, particularly in the hind legs. Although the intact cysts cause no symptoms, when a cyst ruptures it causes myositis. The wild cottontail rabbit is a common host for Sarcocystis, and pet rabbits should be kept away from these rabbits. There is no cure for sarcocystis.

Conditions of Young Rabbits

1. Splay leg

This is a congenital condition and is noticed when the young begin to leave the nest. One or more limbs are spread out, and the rabbit is unable to move properly. Splayed legs can result from femoral luxation or subluxation, distal foreleg curvature, or achondroplasia of the hip or shoulder. There is generally no cure, and euthanasia may the only option. Kits with only one affected limb may be able to manage until they are old enough to undergo amputation of the affected limb, which will greatly increase their overall mobility.

2. Hind leg paresis

This can occur spontaneously in rabbits of 12-14 weeks of age. The hind legs are weak and proprioception and withdrawal reflexes are reduced or absent. The administration of anti-inflammatory drugs, and the oral dosing of tomato juice (a rich source of potassium) appear to resolve the condition in many cases.

3. Musculodystrophy and Hind leg paralysis

These can be seen in association with vitamin E deficiency and can also occur in older rabbits. There is marked skeletal muscle degeneration.

❑❑❑

Chapter – 8

Angora Rabbit Production

Management of Angora rabbit includes important aspects of breeding, feeding, heeding and disease control.

8.1. Reproduction

Angora rabbits of lighter breeds reach breeding age of 6-7 m, medium rabbits at 7-8m and heavier breeds at 8-10 m. Female Angora rabbits do not have a true oestrus cycle with a definite heat period. They may mate at any time of year. Ovulation in does is induced by mating and occurs about 10 -12 hours later to mounting and copulation with the bucks. No doubt there is no definite heat cycle in Angora rabbits, yet during certain periods breeding does are more receptive to bucks than other times of the year. Some of the signs of such receptive period are restlessness, rubbing of chin on the feeding or watering bowls and vulva may be slightly swollen and purplish red.

8.2. Mating

A breeding doe is taken to buck's cage and is allowed to breed. The mating process is normally completed within 1-3 minutes, provide the female is receptive and the male is virile and sexually active. Successful mating is observed when the male after ejaculation falls to one side or backwards emitting a peculiar sound. After mating the breeding doe is taken back to her cage and proper record is made of the mating date so that the exact kindling date is known.

8.3. Gestation

The gestation period of Angora rabbit is 30 – 32days. Two to three days prior to kindling, a nest box is placed in the doe's cage. The nest box should be clean and should contain clean hay, jute or other bedding material.

8.4. Kindling

Prior to kindling, the doe will pull hair from her chest, abdomen and legs to prepare the nest. Kindling takes place usually in the early morning and is completed within 30 minutes. The doe crouches on its hind legs and uses its mouth to pull the kits and eats the foetal membrane in the process. The doe licks and nurses each kit as it is delivered.

On the first day after the litter is born, the nest should be examined carefully to determine the condition of the kits and to remove any dead or deformed infant. The Angoras normally litter 4-8 kits. Most does feed and care for their young when a suitable nest box at about 3 weeks of age and start eating with the dam by nibbling at feed and hay etc. the nest box should preferably be removed at 4 weeks of age but it may be kept for a longer period i.e., 6-7 weeks in cold weather.

8.5. Weaning

Young rabbits may be weaned at 6-8 weeks of age and the doe be rebred after giving a rest of 2 weeks or so. The weaning is done usually by taking away the litter from hutch and leaving the doe alone. After weaning the young rabbits can be fed pelleted feed, hay or some green forage on recommended schedule.

8.6. Sexing

The young rabbit is placed on the forearm with one hand under the rump. The thumb and forefingers of the other hand passes down gently on either side of the sex organ, below the anal opening. In males the penis will protrude out as a rounded tip, while in female rabbits the protruding vulva will appear as a slit opening. The sexes can easily and conveniently be distinguished at the time of weaning.

8.7. Identification

Correct identification of Angora rabbit is essential. Identification can be done in two ways:

1. Individual identification.
2. Cage identification.

8.7.1. Individual identification

Each individual Angora rabbit is allotted a specific number according to its breed, date of birth and serial number. These are to be recorded in their record register. Individual identification is of two types:

(a) Ear tagging

(b) Ear tattooing

8.7.2. Cage identification

Each cage should have a cage card depicting the number, breed, date of birth, mating date, kindling date etc. All these observations are recorded on proper buck and doe performance registers.

8.8. Harvesting of Wool

Harvesting of wool is done mainly by a number of methods as follows:

(a) Shearing

This method of wool harvesting is commonly preferred method in most of the countries including India. The wool matures at approximately 14 weeks but the shearing takes place at intervals of 12-13 weeks. In some cases shearing is done 5 times a year. This may produce wool of shorter length but a greater yield per annum.

Wool shearing is commonly done by an ordinary hair cutting scissors. In this method the animal is handled by the operator with the left hand and wool is clipped with right hand starting with back, shoulders, rump and neck followed by belly, legs, forehead and ears. Some of the wool should be left on the forehead and ears. Some of the wool should be left on the skin depending on the weather conditions to provide thermo-regulatory adjustments in inclement weathers.

In large commercial Angora farms, the shearing of wool is done by electric shearing machines. In this technique the rabbit is shorn by machine. One adult Angora is clipped within 8-10 minutes with this technique. Machine shearing is the best and convenient method as it is quick and does not cause any cut or wound to the rabbit. The main advantages / disadvantages of electric machine shearing are given below:

1. No trouble and pressure on the rabbit shearing without pain to animal.
2. Rabbit can be saved from cold and hot weather by adjusting the wool length left to be clipped.
3. Saves time and labour due to electric machine shearing.
4. Uniform and equal length of fibres at clipping.
5. Convenient at the time of grading.

(b) Depilation

This method of removal of wool from the skin of the rabbit is done by using depilation medicines either by local application or through injections which causes the shedding of wool. This technique is not in practice these days.

(c) Plucking

This method of wool harvesting is mainly adopted in China. In this method small tufts of wool at the hair tips are pulled out using the thumb and the second finger. The plucking is done 4-6 times a year depending upon loosing and moulting. The length of the wool fibre is longest in this method. But tis method is somewhat painful to the animals and takes longer time than other method.

(d) Combing

This method of wool harvesting is mostly followed in France which is called Epilation. It is accomplished by the use of short-handled comb like implement which helps the operator in removing patches of wool. This process is done by fixing the animal on a table by 2-4 tight straps. It is considerably more labour intensive and time consuming. This process is also painful and sometimes causes wounds in the skin.

(e) Chemical method

When mimosin is fed, falling of hair results. In sheep, feeding of excess subabul (*Leucaena leucocephala*) which has mimosin. This causes deflects but it is temporary.

8.9. Angora Wool Quality

There is no normal animal fibre which has the same lightness, softness and purity of fleece as Angora. The weight of Angora wool is 1.1g per cubic centimeter.

Angora wool comprises wooly fibres and the guard hairs. The guard hairs may be classified into two types.

i. Finer types - Has uniserial type medulla of ladder shape.

ii. Coarser type – Has one or two medullary fibres on the base and becomes multiserial at middle.

The composition of Angora fleece is as under:

Type	Average Length (cm)	Average Diameter	
		Head	Tail (μ)
Guard hairs	12-13	90-100	50-60
Wool hairs	8-9	60-80	16-30
Wool	6-7	15-25	10-15

8.10. Wool Grading and Marketing

Different countries and indeed different buyers of the same country have different methods of grading of Angora wool. It is therefore, important to know the grading used by a particular buyer with whom one is dealing. However, the internationally recognized grades of Angora wool are given as under:

I. **Grade I or "A":** Pure white, completely clean, unmated, free from any vegetable matter and fibre length 6 cm and above.

II. **Grade II or "B":** Pure white, free from any foreign matter, non-felted and length between 3 and 6 cm.

III. **Grade III or "C":** White, clean, untangled, free from vegetable matter and length less than 3 cm.

Felted tangled, soiled and stained wool is sometimes un-saleable and if at all sold, it will fetch very low market price. On a well run and organized Angora farm, about 70-80 % wool should be of "A" grade, 20-25%, "B" and "C" grade and rest 3-5% as dirty and matted wool.

❑❑❑

Chapter – 9

Clinical Examination and Laboratory Procedures

9.1. Handling

Rabbits must be handled very carefully. Although they are generally docile they are apt to struggle when picked up, with potentially harmful consequences. All rabbits are prone to vertebral fractures, and this risk is greatest in hutch rabbits or in those on an unbalanced diet that may have concurrent osteoporosis. It is extremely important to support the back and hindquarters whenever a rabbit is handled.

A rabbit should never be picked up by the ears. It can be lifted gently by the scruff of the neck with one hand whilst the other hand supports its bottom (Fig. 9.1), or alternatively lifted with one hand under the chest, and one supporting its hindquarters (Fig. 9.2). The rabbit should be carried against the body to minimize struggling. It is advisable to place the rabbit's carrying box on the floor so that if it does jump as it is picked up it does not hurt itself. Similarly it is usually easier to return the rabbit to the box on the floor at the end of the examination.

The rabbit should be placed on a non-slip surface for the examination. It should be restrained with a hand on its scruff all the time. Some rabbits are calmed if their eyes are covered with the other hand. The majority of the examination can be done in this position; however, it may be easier to hold the rabbit on its back to examine its incisors, and to trim toenails. For this the assistant holds the rabbit against their body with its back supported with one arm, and the hand steadying the outside hind leg. The other hand is cupped over the chest (Fig. 9.3). Alternatively the rabbit can be held by the scruff with the left hand and the back supported by the left arm (Figure 3.4). This frees the other hand for other procedures. Rabbits can be kept calm in this position, and if the handler gently strokes the bridge of the nose whilst talking gently they can be 'tranced' (hypnotised) (Fig. 9.5). Their heart rate and respiration rate do not increase in this position, indicating that they do not find it stressful.

For palpation of the abdomen it may be easiest to hold the rabbit vertically, with its back against the handler's body, and one hand supporting the chest (Fig. 9.6). This frees the other hand to examine the rabbit.

If a rabbit is too fractious or nervous to examine, it may be necessary to sedate it so that the examination can be carried out without injury. Diazepam can be given at a dose of 1±2 mg/kg by intramuscular injection.

9.2. Examination

The examination should be conducted in a logical manner so that each system is investigated. Beginning at the head, the incisors should be inspected for enamel quality and signs of malocclusion. The molars are harder to inspect: an aural or vaginal speculum will help visualise these back teeth; however, if dental problems are suspected a fuller examination will be required under sedation, as it is possible to miss lesions with a speculum. A wet chin, or dampness on the medial aspect of the forelegs will indicate that the rabbit has been dribbling through dental discomfort.

The lower jaw should be felt for any bony abnormalities, or abscesses that may be associated with the tooth roots. The head should be held level, and the eyes should be examined for even pupil size and other ocular abnormalities. There should be no ocular discharge, and pressure at the medial canthus of the eye will reveal any purulent material in the tear duct. The nares should be examined for the presence of a nasal discharge, and again the medial aspect of the forelegs may be matted with discharge as the rabbit wipes its nose. The ears should be examined for the presence of exudate. Normal ear wax is golden in colour, but a dry crusty exudate is seen with ear mites. The thorax should be auscultated. Interpretation of chest sounds requires experience, as referred upper airway sounds are normal because the rabbit breaths through its nose.

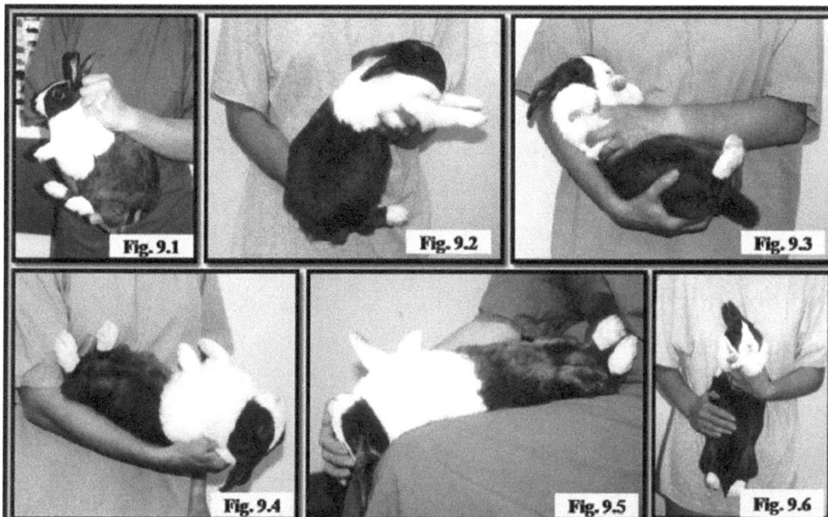

Fig. 9.1 : Rabbit being held correctly with one hand holding scruff of neck while the other hand supports the hindquarters. **Fig. 9.2** : An alternative way of correctly holding a rabbit is to place one hand under the chest and support the hindquarters with the other hand. **Fig. 9.3** Rabbits can also be held with their back supported by one arm and one hand steadying the outside hind leg, with the second hand cupped over the chest. **Fig. 9.4** : The rabbit can be held by the scruff of the neck with one hand, and with the back supported by the other arm. **Fig. 9.5** : Rabbits can be tranced if the handler gently strokes the bridge of the nose while talking gently. **Fig. 9.6** For palpation of the abdomen, it is easier to hold the rabbit vertically with its back against the handler's body, and one hand supporting the chest.

Many rabbits also have a degree of dry, adventitious noise associated with previous low grade respiratory infections. The heart rate is usually 120±250 beats/minute. Rabbits in severe respiratory distress will have an arrhythmic heart rate (above 200 beats/minute) and cyanosed mucous membranes. Abdominal pain may present as respiratory distress.

The abdomen should be palpated gently because the organs can bruise and haemorrhage easily. The presence of gastric bloat, gastric ileus and caecal ileus can be detected on examination. The abdomen can also be auscultated for the presence of gut sounds: an absence of sounds will indicate ileus.

Abnormalities of the uterus such as uterine adenocarcinoma may be palpable. The presence of growths or cysts in the mammary tissue may indicate this as a likely diagnosis. The bladder should be palpated very gently, as it is thin-walled and easily ruptured. Further investigation of the bladder and other abdominal organs may require radiography. The kidneys may be easiest to palpate with the rabbit in dorsal recumbency: the left kidney is very mobile.

The limbs should be examined for abnormalities. The plantar aspect of the hocks is a frequent site of inflammation and haemorrhage (pododermatitis). Rabbits that are presented because the owner has found some blood in the cage should have their toenails and hocks examined carefully, together with the routine genitourinary investigation.

Examination of the skin includes a search for wounds, abscesses and mite infestations. The perineal area and dorsum should be examined closely for blowfly myiasis, especially in any rabbit that is 'off colour' in warm humid weather.

The rabbit should be weighed. This is particularly important if any medication is to be administered. Paediatric scales are ideal for this purpose (Fig. 9.7). The rectal temperature can be taken at the clinician's own judgement, as the findings are often not helpful in making a diagnosis. The normal rabbit temperature is 38.5±408C (101.3±1048F); however, this can increase with stress or during a period in a carrier in the waiting room. Conversely sick animals generally have a low body temperature.

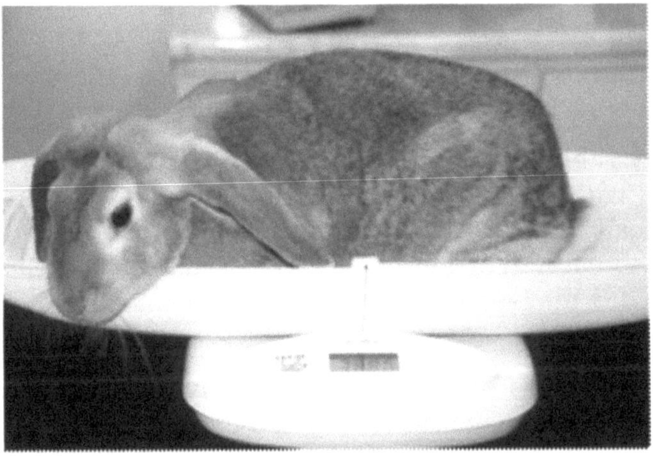

Fig.9.7: Paediatric scales, ideal for weighing rabbits.

9.3. Oral Medication

Oral medication may be formulated as a liquid or tablets. Liquids can be easily syringed into the mouth, or, where appropriate added to the drinking water. Tablets can be mixed with a favourite food such as banana, or mixed with strawberry jam. Debilitated rabbits that may have a nasogastric tube in place can be given liquid medication through this route.

9.4. Injection Techniques

Small gauge needles (22±27 gauge) should be used for injections to minimize discomfort. For accurate dosing an insulin syringe may be recommended. Subcutaneous injections are given in the scruff of the neck. This site is also suitable for the administration of subcutaneous fluids as rabbits have a large subcuticular space under the neck and shoulders.

Intramuscular injections can be given in the cranial aspect of the quadriceps muscle. Only small volumes should be administered in this way (up to 1.5 ml in a large rabbit).

Intravenous injections can be given into the cephalic or lateral saphenous veins. Care should be used if the marginal ear veins are used, as many drugs are irritant and may cause sloughing of the pinna.

9.5. Catheterisation

Intravenous catheters can be placed in the cephalic or lateral saphenous veins for the administration of fluids. In the conscious patient the cephalic vein is preferable, as it is less affected by the rabbit moving its legs. The catheter should be bandaged on well to prevent the rabbit removing it.

Intraosseous catheterisation is required for very debilitated rabbits for whom intravenous catheterisation may be impossible. Ideally this technique should be practised on some cadavers first. The femur is the most frequently used site, although catheters can be placed in the humerus or tibia. The procedure is painful and should be done under anaesthesia, or local anaesthetic. A 20±22 gauge spinal or hypodermic needle can be used; with the latter a stylet can be made using sterile surgical wire. The placement of the catheter should be done under sterile surgical conditions. The area over the head of the femur is clipped up, a small incision is made over the proximal aspect of the greater trochanter, and the needle is introduced through the trochanteric fossa into the marrow cavity. 3±5 ml of sterile saline can be injected slowly to test the patency of the catheterisation. The needle should be flushed with heparinised saline and capped. Any medication and fluids should be administered very slowly through this route. Parental antibiotics should be used while the catheter is in place, and for a few days after.

9.6. Pre-Anaesthetic Preparation

Rabbits should not be starved before surgery. They are unable to vomit so there is no risk of aspiration of stomach contents. A period of starvation will cause hepatic lipidosis, and will delay the onset of eating after surgery. Post-anaesthetic gastric ileus is also more common. Stress should be minimised by the administration of some premedication.

If the patient is dehydrated fluids should be given intravenously or subcutaneously before surgery. Hypoglycaemia is a potential problem in any rabbit, particularly if its appetite has been reduced, and lactated Ringer's solution or glucose±saline can be given: up to 80±100 ml can be given at one time. A sick rabbit should be stabilised as well as possible before being anaesthetised; if fluids have been given subcutaneously they will take 6±8 hours to be fully adsorbed, and if possible anaesthesia should be delayed until the rabbit is hydrated.

9.7. Premedication

The use of premedication will reduce the stress of anaesthetic induction, and will also allow the rabbit to be maintained on a lower percentage of inhalation anaesthesia. Some premedication drugs are good analgesics, and will give post-operative analgesia as well.

Acepromazine (ACP) can be given at a dose of 0.1±0.5 mg/kg by subcutaneous injection. It is not analgesic, and causes vasodilation. This facilitates subsequent intravenous injections and blood collection. ACP can be combined with buprenorphine to provide sedation and analgesia. ACP at a dose of 0.1 mg/kg and buprenorphine at a dose of 0.01 mg/kg can be combined in the same syringe and given by subcutaneous or intramuscular injection.

Butorphanol can be given at a dose of 0.1±0.5 mg/kg by subcutaneous injection. This will also provide post-operative analgesia of 3±4 hours' duration and is particularly useful for ovariohysterectomy cases.

Buprenorphine can be given at a dose of 0.01±0.05 mg/kg by subcutaneous injection.This will also provide post-anaesthetic analgesia of 6±12 hours' duration.

Diazepam can be given at a dose of 2 mg/kg by intramuscular injection.

9.8. Sedation

Some procedures such as dental work, tear duct flushes and the lancing of abscesses can be done under sedation alone. Various drug combinations are suitable. Combinations that include either xylazine, medetomidine or butorphanol will also have an analgesic effect. The rabbit must be constantly observed as the sedation takes effect to avoid it adopting a position that could compromise its airway. Close monitoring is also important during recovery for the same reason. Most of these sedatives cause hypoxia, and the administration of oxygen via a face mask or endotracheal tube during the period of sedation is recommended.

Xylazine at a dose of 5 mg/kg and ketamine at a dose of 25±35 mg/kg can be mixed in the same syringe and given by intramuscular injection. Sedation is achieved in 5±10 minutes, and full recovery may take several hours, during which time the patient should be kept warm and turned regularly.

Ketamine at a dose of 15 mg/kg and medetomidine at a dose of 0.25 mg/kg can be mixed in the same syringe and given by intramuscular or subcutaneous injection. This combination can be reversed with atipamezole at a dose of 1 mg/kg given by intramuscular or subcutaneous injection. Recovery from ketamine and medetomidine alone may take 2 hours; if reversed with atipamezole recovery is even quicker.

Ketamine (30±40 mg/kg) and diazepam (2 mg/kg) can be given in combination by intramuscular injection.

Sedation and analgesia can be provided with the combination of ketamine (10 mg/kg), medetomidine (0.2 mg/kg) and butorphanol (0.1 mg/kg) given by intramuscular injection. This can be reversed with 1 mg/kg atipamezole by intramuscular injection without losing the analgesic effects.

Fentanyl/fluanisone (Hypnorm, Janssen Animal Health) can be given at a dose of 0.3 ml/kg by intramuscular injection in combination with diazepam (1 mg/kg) or midazolam (2 mg/kg) by intraperitoneal injection.

Surgical anaesthesia and muscle relaxation lasts for 20±40 minutes. This sedation can be partially reversed with buprenorphine (0.01±0.05 mg/kg) or butorphanol (0.1 mg/kg) by subcutaneous injection.

9.9. Inhalation Anaesthesia

Gaseous anaesthetics can be administered with oxygen. Nitrous oxide should not be used in the rabbit as it increases the tendency of the gastrointestinal tract to bloat.

Isoflurane is the anaesthetic of choice for rabbits. It is not as noxious as halothane and there is less of a tendency to breath hold during induction. Induction can be done with a concentration of 4%, and anaesthesia can be maintained with 1±2%. Lower concentrations are possible if the rabbit has received premedication. As 98% of the isoflurane is expelled through the lungs, recovery is rapid and no stress is placed on other organs such as the liver and kidneys.

Halothane can be used to induce anaesthesia at a concentration of 4±5% and anaesthesia can be maintained with 2%, or lower if a premedicant drug has been given. Care must be taken as halothane sensitises the myocardium to the arrhythmic effects of adrenaline (released in the stress of handling).

9.10. Induction

The procedure of induction must be as stress-free as possible. Rabbits that have been sedated will be more relaxed. Rabbits have a tendency to breathhold if stressed, and this is accompanied by bradycardia, hypercapnia and hypoxia.

The author first trances the patient by holding it upside-down and stroking it along the bridge of the nose. When the rabbit is relaxed oxygen is introduced through a face mask. After the rabbit has breathed oxygen for 2±3 minutes the anaesthetic gas is introduced, and induction proceeds smoothly (Fig. 9.8). Care should be taken not to hold the rabbit tightly around the chest during restraint.

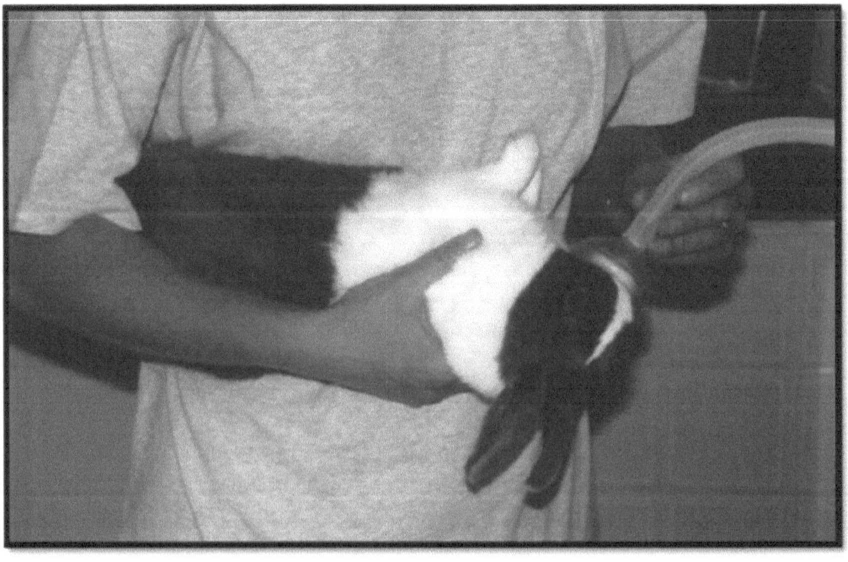

Fig.9.8: Oxygen and anaesthetic gas can be introduced through a face mask. Care is taken not to hold the rabbit tightly around the chest.

9.11. Maintenance of Anaesthesia

Anaesthesia can be maintained via a face mask, or endotracheal tube. The placing of an endotracheal tube takes experience, and after several failed attempts a face mask should be used, as rabbits are prone to laryngeal oedema. With a good scavenging system for the waste anaesthetic gas a face mask is sufficient for short procedures.

The rabbit should be kept on a heat pad during surgery, and its temperature monitored by rectal thermometer. The rabbit's temperature can drop dramatically, particularly during abdominal surgery, and even during minor procedures rabbits have a tendency to become hypothermic.

9.12. Endotracheal Intubation

Endotracheal intubation can be difficult in rabbits, particularly the smaller breeds. The shape of the rabbit's mouth makes it difficult to open it wide to visualise the back of the throat. Small rabbits require a 2mm endotracheal tube; larger rabbits may require a 4mm tube. Intubation can be done under direct visualisation of the larynx, or blind. Intubation is generally only possible if the rabbit is heavily sedated. Several failed attempts at intubation may cause laryngeal haemorrhage and oedema, and after three attempts a face mask should be used instead, as rabbits with traumatised airways do poorly post-operatively.

For direct visualisation the rabbit is placed in dorsal recumbency and the larynx viewed with an otoscope. The larynx can be sprayed with lignocaine to reduce the risk of laryngospasm. After visualising the larynx, an introducer is passed into the trachea (a urinary catheter can be used for this) and the otoscope removed. The endotracheal tube can the be passed over the introducer into the trachea. The rabbit may cough as the tube enters the trachea indicating its correct placement.

For intubation without visualisation of the larynx, the rabbit is held in sternal recumbency with its head stretched up. The endotracheal tube is advanced to the larynx and observed carefully. The presence of condensation in the tube, or audible breath sounds through the tube, indicate that the tube is close to the larynx. As the rabbit breathes in the tube is advanced. The continued presence of breath sounds indicates that the tube is in place. The rabbit may also cough as the tube enters the trachea.

9.13. Anaesthetic Monitoring

Monitoring of the patient is important during surgery. The rectal temperature should be frequently checked as rabbits have a tendency to hypothermia. The depth of anaesthesia can be monitored by the ear pinch and hind-leg withdrawal reflex which should be just present. The foreleg withdrawal disappears later, and if this is absent the anaesthetic must be lightened. Unlike dog and cat patients, the position of the eye during anaesthesia is not significant and is not helpful in monitoring the depth of anaesthesia. Chewing is a sign of light anaesthesia, and may be associated with a rapid awakening from the anaesthesia. Heart and respiration rates should also be monitored.

9.14. Post-Operative Care

The rabbit should be allowed to recover in a warm peaceful environment. If it is one of a pair-bond its partner should be introduced as soon as possible. Food and water should be offered, particularly tempting greens such as kale and parsley.

Sutures are best protected with light dressings. Rabbits do not tolerate Elizabethan collars well; they are unable to practice coprophagy and may become anorexic and depressed.

Post-operative analgesia should be given unless analgesia was given preoperatively, or as part of a sedative combination.

- Aspirin (100 mg/kg) orally every 12±24 hours.
- Butorphanol (0.1 mg/kg) by subcutaneous injection every 8 hours.
- Buprenorphine (0.01±0.05 mg/kg) by subcutaneous injection every 8± 12 hours.
- Carprophen (1.5±2 mg/kg) given orally or by subcutaneous injection every 12±24 hours.
- Flunixin (1.1 mg/kg) by subcutaneous injection every 24 hours.
- Ketoprofen (1±3 mg/kg) by subcutaneous injection every 24 hours.
- Meloxicam (0.1±0.2 mg/kg) orally every 24 hours.

9.15. Surgical Procedures

Descriptions of specific surgery are covered in detail in relevant places in the text. However, a few general considerations should be mentioned. The rabbit has a great tendency to form post-surgical adhesions, particularly after abdominal surgery. It is possible that the talcum powder on sterile surgeon's gloves may trigger adhesion formation, and it is important to wear gloves without powder inside, or wash the powder off the gloves. Adhesion formation is more common with the use of chromic catgut. Although this suture material is acceptable for routine castrations and ovariohysterectomies, a synthetic absorbable suture should be chosen for more complicated abdominal surgery.

If there is concern about post-surgical adhesions, verapmil can be given by intraperitoneal injection at a dose of 0.2 mg/kg every 8 hours for nine doses.

Rabbits are very good at removing their own skin sutures. However, they do not tolerate Elizabethan collars (the standard method of preventing other companion animals from interfering with their sutures); they become depressed and are unable to practice coprophagy.

Alternatives to the simple interrupted skin suture are subcuticular sutures with a synthetic absorbable suture, a continuous close apposition (interlocking) suture with synthetic non-absorbable material, or skin staples. Skin staples are generally well tolerated; however, there is still a small risk that they could be removed, and worse, ingested by the rabbit. The author prefers the continuous interlocking suture. It is a quick and easy method of wound closure, and appears difficult for the rabbit to undo. Even if part of the suture gets bitten through, the interlocking nature of the suture means that the wound remains in apposition to allow healing to take place.

❑❑❑

Chapter – 10

Rabbit Behaviour

Behaviour have an influence on rabbit housing. Since the domestication of the rabbit is recent in terms of species evolution (200 to 300 generations at most) the behaviour of the domestic rabbit is still much like that of the wild rabbit. The reactions of wild rabbits will often provide explanations for the problems of housing domestic rabbits and suggest ways of solving them. Common behaviours observed in rabbits are:

1. Territorial behavior.
2. Social behavior.
3. Sexual behavior.
4. Maternal behavior.
5. Feeding behavior.
6. Aggressive behavior.

10.1. Territorial Behavior

Rabbits live in sedentary fashion in a territory the size of which depends on the conditions of food supply. They mark their territory, their fellows and their offspring with the aid of a gland found in hair follicles under the chin. The bucks also mark off their territory with their urine. The rabbits dig burrows in which they take shelter at the slightest sign of warning. There they live in a "society". Before parturition, however, the doe digs a special burrow where her young are born and where she returns once a day to nurse them. This is why domestic rabbits should have durable living quarters, providing either a refuge from disturbances or a peaceful environment that makes a refuge unnecessary. Any new and sudden change (noise, presence, smell) will make the first rabbit in the group to notice the disquieting novelty thump his hind foot to warn his fellows of danger. To prevent panic in the rabbitry the breeder should take care that changes that might upset the animals are avoided. When a rabbit is put in a new cage he will explore it and then mark it with his smell. The more strange odours there are in the cage the longer this task will take.

10.2. Social Behavior

Wild rabbits live in colonies in which females outnumber males. Each female, with or without offspring, attacks the young of other does. Bucks act as moderators at this stage. When the young males reach puberty, however, the adult males try to eliminate them as rivals by castrating them. Such conflicts can be prevented by isolating each adult rabbit in an individual cage. Before puberty, young rabbits can be reared in groups. Attempts to rear breeding animals in groups are bound to fail because the does are so aggressive towards the young, especially when the animal's living space is cramped. Females without young can be reared in groups provided each female has at least half a square metre of space for herself.

10.3. Sexual Behavior

Ovulation in the does is brought on by mating, so one might expect mating to be possible on a quasi-permanent basis. The buck is so very territorial that when he is put in a female's cage his first act is to mark this new territory with his smell, while the doe tries to eliminate the intruder. But if a doe is put in a buck's cage the immediate reaction of both animals is sexual. For a receptive doe preparation for mating takes 20 to 120 seconds, and the act itself less than a second. For servicing, therefore, it is the doe that should be moved. This is relatively easy because does are calmer and weigh less than bucks (3 to 6 kg). For mating to be supervised, the animals need to be visible in all parts of the cage. Access to the buck's cage should be simple so that the does can be easily introduced and removed.

10.4. Maternal Behavior

Before kindling, the doe makes a nest with various materials plus fur that she pulls from her abdomen. The wild doe's nest is made at the end of the private burrow she digs for kindling. The domestic rabbit does not usually have the opportunity to do this, so a private area should be set apart for her. It is preferable to provide her with a nest box that approximates the natural burrow. A box like this is useful in a farm rabbitry and essential in wire cage production. After the young are born (six to 12 per litter), the doe nurses them once every 24 hours for about a month. To allow the motor coordination and heat-regulation capacity of the baby rabbits to develop, the nesting box should be maintained for at least two weeks. It should be big enough to accommodate the doe and her litter during nursing.

10.5. Feeding Behavior

Rabbits will drink and eat at any time in the 24 hours, although they tend to feed nocturnally. Feed and water should therefore be available over periods of several hours, whether feed is rationed or ad lib. The feed must not be allowed to get dirty, which is inevitable if it is strewn on the ground. From the age of three weeks young rabbits begin to eat the same feed as the doe. Their small size allows them to slip easily into forage racks or dry feed hoppers, so this equipment must be designed to keep them out. A drinker and feeding rack should be provided for each cage. The animal must be able to reach the feeders and fodder racks, as must the caretaker to top them up frequently. An automatic or semi-automatic drinker is easy to make, however.

10.6. Aggressive Behavior

One of the commonest reasons that rabbits are taken to rabbit sanctuaries for rehoming is because they have become aggressive. If the reason for the aggression can be determined then the behaviour can usually be modified. Common aggressions observed in rabbits are:

a) Territorial aggression

Rabbits that attack their owners when they are in their hutch are thought to display territorial aggression. This may occur seasonally, such as seen with entire does as they reach sexual maturity and during the breeding season, or as the result of pseudopregnancy. For these does neutering will calm their behaviour. Rabbits that attack hands that approach them as they sit in their hutch may actually be exhibiting fear aggression towards the hand, rather than territorial aggression since rabbits are fearful of hands that approach from their sides or face, and of hands that smell unusual. The hand should always approach from above or behind the head, and often the living quarters can be adapted to incorporate a top opening door to make this possible.

b) Food aggression

Rabbits may defend their food bowl, and attack the hand that tries to remove it. In the wild, rabbit is accustomed to a poor diet, and if it finds an area of lush grass, it is its instinct to protect it. Rabbits that guard one food bowl in the same manner should be given many sources of food around their environment, so that one single site does not become so important. Fresh food can be scattered around, allowing the rabbit to develop its natural instinct to forage for food. Rabbits which are provided with a natural high fibre diet of hay and leafy vegetables that takes them several hours to forage for and graze on are likely to be more satisfied than the rabbits which receive their daily ration as a bowl of concentrated food which they can eat in less than half an hour. Diets that are rich in protein and carbohydrate (concentrate ration) are in themselves likely to cause behavior problems. This is equivalent to feeding enough protein and carbohydrate to a racehorse, and then expecting it to be content to stand in a stable all day.

c) Fear aggression

The rabbit is a prey animal, and is naturally fearful of predators, and in particular of being chased, caught and picked up. When a rabbit is scared it can let out a high pitched scream. In the wild, this scream is thought to warn other members of the colony, whilst attempting to frighten the predator. In any situation, if a rabbit shows aggression, it is important to analyze the events to determine if the aggression is a fear response, and then the behavior can be modified by removal of the fearful stimulus. Rabbits rely heavily on their hearing to give them early warning signals from the environment. Deafness is a common cause of fear, and fear aggression. This may be due to an inner ear infection, or more commonly a heavy build up of wax and pain associated with an ear mite infection. The ears should be examined and treated accordingly.

d) Sexual aggression

A rabbit can become aggressive once it reaches sexual maturity. This aggression may be towards other companions, or towards its owner. In does, this aggression may be cyclical, relating to nest building and pseudopregnancy. Does with litters may show aggression when protecting their young, or a doe may protect other submissive companions. Does that may have uterine hyperplasia or adenocarcinoma can become aggressive. Aggression related to sexual maturity will greatly improve after neutering.

e) Pain aggression

Rabbits that are in pain will become short-tempered and aggressive. This pain may be from any source, and must be investigated accordingly.

❑❑❑

CHAPTER - 11

Slaughter and Skinning of Rabbits

Important by-products from rabbits are meat, wool, planner and skin. It can be made highly profitable if products are wasted because people do not have enough knowledge about their processing.

11.1. Slaughtering and Dressing

1. Rabbits are killed by one of two methods. The preferred method is dislocation of the neck. The rabbit is held firmly by the rear legs and head; it is stretched full length. Then with a hard, sharp pull, the head is bent backward to dislocate the neck. The rabbit can also be struck a hard, quick blow to the skull behind the ears. A blunt stick or side of the hand is commonly used to incapacitate the rabbit. Both methods quickly render the rabbit unconscious.
2. After dislocation or stunning, the rabbit is hung by one of the hind legs above the hock joint. The head is immediately removed to allow complete bleeding. The forefeet are then removed.
3. The next step is to cut the skin around the hock joints of the legs and then to cut between these points across the lower part of the body. Remove the tail and pull the skin down and forward over the body. The skins of young or fryer age rabbits are easily removed in this way; it is more difficult to remove the skins of older rabbits.
4. After the head, forefeet and skin are removed, the carcass, while still hanging, is opened to remove the viscera. Make a

cut from the lower part of the abdomen near the anus to the mid-point of the lowest rib. The intestinal tract and lungs are normally removed. Liver, kidneys, and heart remain with the carcass.

5. Remove the carcass from the hanger and cut off the rear feet at the hock joint.
6. Wash the carcass with clean, cold water to remove hair and any other soil or debris, and store it at a cold temperature, preferably at 35°F and not over 40°F.
7. Do not hold dressed carcasses for any length of time in water as they absorb excess moisture which becomes considered as a contaminant.

When large numbers of rabbits are to be slaughtered, the operation on an assembly line basis is much more rapid and efficient. Dressed rabbits may be sold whole or can be cut into parts. Various types of marketing devices (plastic cartons and trays) are available for packaging one or more carcasses.

11.2. Collection, Preservation and Storage of Pelts

11.2.1. Skinning

Skinning should be carried out in a manner that ensures the largest possible skin surface, which is an important part of its value. The first cut is usually an incision at the hind feet, passing from one thigh to the other. The skin is then pulled off. The skin on the head is of no commercial value but it is preferable to keep it because it allows better stretching. This operation should be done with care to avoid mutilation, knife marks, grease (which oxidizes and burns the skin) or bloodstains. All these defects reduce the value of the pelt, especially when the coat is originally of good quality. The sequence of skinning operations is illustrated in Fig. 11.1.

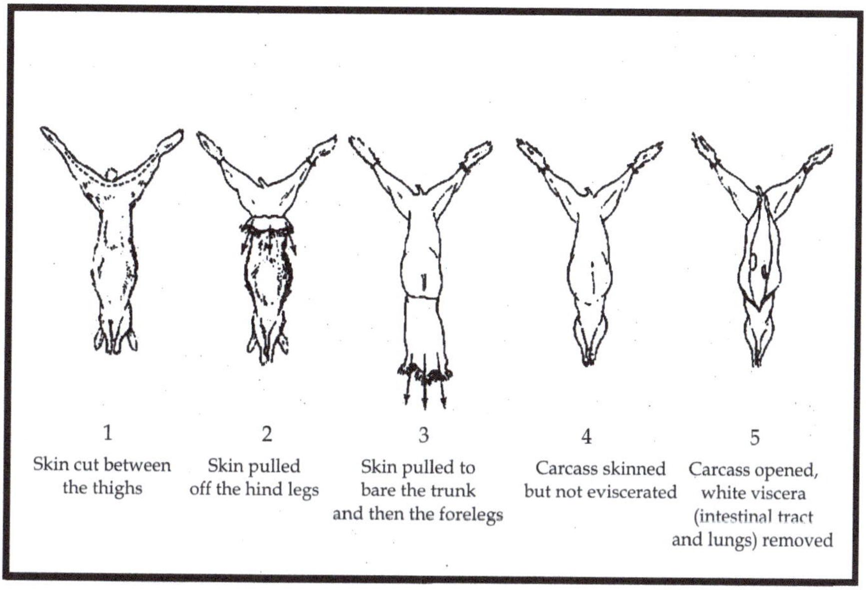

Fig. 11.1: Skinning a rabbit

11.2.2. Preservation

Rabbit pelts are preserved by drying. This is a simple operation which can be done anywhere and costs little. Drying should start immediately after the skin has been removed. It must cool off quickly and dry out to prevent the action of enzymes in the derma which attack the hair root and cause the hair to fall. If fresh pelts are left in a pile for even a short time (more than 15 minutes) a rapid bacterial fermentation will set in and cause the hair to fall out in patches. Many pelts are lost this way through lack of elementary care.

The skins are shaped on a frame. They should not be excessively stretched, nor should there be any creases. The frame can be a board or a steel wire frame. Straw should not be used as padding as it can deform the pelt in places.

During drying, air should circulate freely and the skins should not come into contact with one another. It is unacceptable to accelerate drying by exposing the skins to the sun or to hot air; above 50°C the collagen of the derma is altered irreversibly and the skin cannot be processed. They should be dried in the shade or in the dark in a well aired dry place (optimum temperature 18° to 22°C). Twenty-four hours later it is best to remove fatty deposits on the shoulders and belly to avoid local hotspots.

11.2.3. Packaging and storage

The pelts are arranged in piles when they are perfectly dry in a cool airy room, with insecticide (naphthaline) between each layer of skin. It is best to grade the pelts without delay, the grading being more or less elaborate according to the size of the stock in question. At least the different qualities should be separated immediately and the white pelts from the coloured.

Whether the destination of the pelt is fur or hair production, all operations from skinning to storage must be carried out with care and attention. The slightest fault in handling results in a lowering of grade, which is all the more serious when a high quality skin is involved and all the work carried out previously is lost. The greater the homogeneity and quality of the pelts the more attractive they will be to the trader, which is particularly important at times of market depression.

11.3. Curing and Glossing of Pelts

11.3.1. Curing

Processing the pelt to the semi-finished stage requires a series of operations:

1. Dipping: Rehydration of the pelts with water, salt and possibly soap, followed by rinsing.
2. Fleshing: The rabbit skin has a peculiarity, a thin collagenous film on the flesh side. This membrane, which is impervious to curing products, should be removed. This is a delicate, labour-intensive operation, carried out on the rehydrated skins.
3. Dressing: The special tanning for rabbit skins generally uses a specific blend of salt, alum and formol.
4. Thirming: It is necessary to thin down the thicker skins. This is highly specialized work, demanding great precision to avoid holes in the skin, cutting hair follicles and causing hair loss. A second dressing is carried out on the thirmed skins.
5. Greasing: Nourishing the skin with oil. This operation is labour-intensive; finishing: this gives the skin a pleasing appearance and consists of removing grease (stirring in a tub with absorbents), beating (tossing in a meshed cylinder to remove absorbents, sawdust, grit, kaolin) and lifting the hair to set it in place. Machines can be used for all three finishing stages.

11.3.2. Glossing

This is a complicated finishing operation, with variations such as shaving or colouring according to the final product required. It calls for much handling, expertise and imagination (mixing of dyes, special effects, etc.). It is often the furrier who, having chosen a lot of rough furs, decides on the final appearance they will be given. For a coat, 20 to 30 skins will be needed. The making up of "bodies" (remnants of fur sewn together and sold by length), which is labour-intensive and not highly automated, can be done in developing countries or in countries where the labour is less expensive.

11.4. Sorting and Grading Pelts

In an unsorted batch of rabbit skins valuable pelts can be found side by side with useless waste, so sorting and grading should be done as early as possible.

11.4.1. Sorting

Sorting, the first operation, determines the future use of the skin. Skins are sorted into three grades:

1. Pelts for dressing (the term "dressing" instead of "tanning" is used for fur): These are the best skins, with regular shape, intact, homogeneous, dense, a well-formed coat, a flawless skin. Their price may be 20 times that of unsorted quality skins.
2. Pelts for shorn hair: These usually lack the proper shape or are not homogeneous enough for fur products. The hair, however, is sufficiently long and healthy. It is therefore machine shorn and used for textiles or felting.The skin is cut into fine strips (vermicelli) and made into glue or fertilizer. This technique allows much of the pelt to be recycled.
3. Waste: Waste is unusable except for fertilizer (the hair is gnawed, cut, soiled, sweaty, parasite-ridden). Such skins push up the costs of labour, processing and transport.

11.4.2. Grading

The following grading system is used for pels:

- For pelts for shorn hair:

Rejects	Hair weight 10 to 18 percent of the dry pelt weight
Ordinary	Hair weight more than 18 percent of dry pelt weight
Good quality	With guard hair removed for glove-making

- For fur pelts:

Grading is more complex for fur pelts, as colour, size and quality are all considered. The colours are white, range of grey, range of red (nankin), mixed and black. Size is assessed by weight per 100 dry pelts as shown in the table below:

Entre-deux	12 to 13 kg / 100 pelts (100 to 140 g per pelt)
Cage	13 to 20 kg / 100 pelts (150 to 210 g per pelt)
Heavy	26 to 40 kg / 100 pelts (250 to 350 g per pelt).

❑❑❑

Annexures

ANNEXURE-I

Facts about Rabbits

1. Rabbits are classified as "Lagomorphs".
2. Baby Bunnies are commonly referred to as "Kittens".
3. A Rabbit's teeth never stop growing.
4. Female rabbits are referred to as "Does".
5. Male rabbits are referred to as "Bucks".
6. In the wild rabbits live in groups called "Warrens".
7. Act of parturition is known as Kindling.
8. Today's domestic rabbit is descended from the European Rabbit.
9. They are pseudoruminants.
10. Rabbits do not vomit.
11. The backbone of a rabbit is very fragile and can break easily when handled improperly or dropped on its back.
12. Domestic rabbits are born without fur.
13. Bunny Rabbits are generally the most active during the evening and early morning.
14. When rabbits grind their teeth it can sound like a purr.
15. Rabbits are nearsighted.
16. Rabbits are not rodents but can resemble them in appearance.
17. The largest number of kits every born in a litter is 24.
18. A rabbit has five toe nails on its front two paws and four toe nails on its back two feet (18 toe nails per rabbit).

19. The longest rabbit ears ever recorded measured up to be 31.125 inches long.
20. The world's heaviest rabbit is Darius, weighing an unprecedented 50 pounds.
21. Hundreds of years ago rabbits were often released on deserted islands in hopes of giving ship wrecked sailors a reliable food source.
22. Some places in the world have had serious trouble with rabbit overpopulation.
23. There are over 150 recognized rabbit coat colors and varieties.
24. The current world record for a rabbit long jump is 3 meters.
25. The current world record for the rabbit high jump is 1 meter.
26. Rabbits only sweat on the pads of their feet.
27. A rabbit will eat its own cecotropes (night droppings)- they are a valuable source of protein.
28. A group of bunnies from the same mother are called a "Litter".
29. Domestic rabbit kits are born with their eyes shut.
30. It has been estimated that in Australia rabbits destroy around $600 million worth of crops each year.
31. Throughout history the rabbit has been seen as a symbol for fertility.
32. Not all rabbits will "Breed like rabbits".
33. In ancient Egypt rabbits were used as sport for dog racing.
34. The scientific name for the rabbit is *Oryctolagus cuniculus.*
35. Rabbits and hares commonly found in the United States of America include the cottontail, jackrabbit, snowshoe rabbit and the domestic rabbit.
36. Rabbits can interbreed with Hares.
37. Pet rabbits tend to live to be much older that wild rabbits.
38. Popular rabbit organizations include; Rabbit Empire, American Rabbit Breeders Association, House Rabbit Society and the British Rabbit Council.
39. Rabbits were an important home meat supply during World War II.

40. Some rabbits are raised specifically for their fur.
41. A rabbit business is commonly called a "Rabbitry".
42. Many bunnies are confined to a rabbit cage environment.
43. Millions of pounds of rabbit meat are consumed each and every year.
44. North America has the largest domestic rabbit population.
45. In England rabbits have become the third most popular pet (after dogs and cats).
46. Some people raise bunnies specifically for photography subjects.
47. There are around 180 different rabbit breeds worldwide.

ANNEXURE-II

Body Parts of a Rabbit

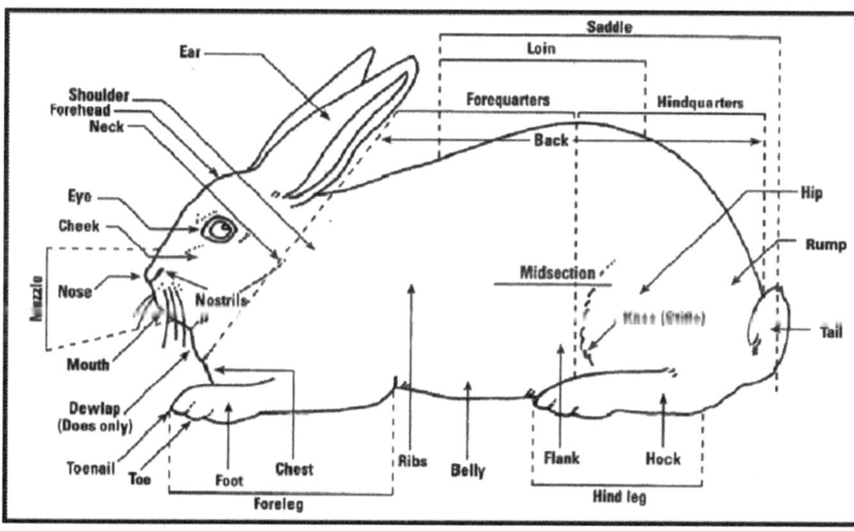

ANNEXURE-III

Model Project for Rabbit Farming (Meat Production)

Calculation Scale

(a)	Male and female ratio of rabbits	=	1 - 2
(b)	Age of breeding	=	100 – 110 days
(c)	Weight of breeding buck	=	3 kg (Approx.)
(d)	Marketing age	=	100 days (Approx.)
(e)	Feed requird		
	a. Upto 15 days	=	Mother's milk only
	b. 16 – 42 days	=	Plenty of grasses
	c. 43 – 100 days 30 – 60 g feed	=	Grasses along with
	d. Adult 60 – 100 g feed	=	Grasses along with

A. Feed Capital Required

a. Construction of cages for adult @ Rs. 300/- for 2 cages = Rs. 600

b. Construction of cages for breeding stock = Rs. 500

c. Cost of rabbit @Rs. 60/- for 3 rabbits = Rs. 180

d. Cost of feeder and waterer = Rs. 100

e. Cost of insurance @Rs. 6% of rabbit cost = Rs. 10 .

Total = Rs. 1390

B. Recurring Expenditure

a. Cost of feed for 180 days @80g per rabbit = Rs. 390 costing Rs. 9/- per kg

b. Cost of feed for 60 days @20g for young stock = Rs. 216

c. Cost of medicines = Rs. 50

Total = Rs. 656

Total expenditure A+B = Rs. 1,390 + 656 = **Rs. 2046**

C. Expected Income

a. Income by selling 20 rabbits with body weight of 2kg = Rs. 2400 @ Rs. 35/- per kg

b. Income from empty gunny bag Rs. 10 for 3 bags = Rs. 30

Total = Rs. 1430

Net income = Rs 1430 - Rs 656 = **Rs. 774**

ANNEXURE-IV

Difference between Wild & Domesticated Rabbits

S. No.	Wild rabbit	Domestic rabbit
1.	Usually grey/greyish brown in colour.	Developed and present in variety of colours.
2.	The young called bunny/kit are born with fur with open eyes and can run about almost immediately.	Born hairless with eyes closed. Open their eyes after 10 days and cannot run immediately.
3.	The mother does not build nest but gives birth to young ones in a reasonably sheltered manner.	Mother needs confinement and builds a well hidden warm nest for its bunnies.
4.	Has long legs, can jump and leap 15 - 20 feet in one bound.	Has short legs and can run and hop but cannot leap.
5.	Have long and upright ears.	Have comparatively short ears.
6.	Much heavier in body weight.	Generally lighter, but with improved methods heavier domesticated rabbits have also been produced.
7.	Fight among themselves by kicking with their powerful hind (back) legs and biting.	Very docile comparatively and are more popular house pets. Males sometimes fight by scratching with their fore-legs and biting

ANNEXURE – V

Recommended nutrient levels in diets for various categories of rabbits reared intensively

Dietary composition (assuming a dry matter content of 89%)	Units	Growing rabbits (4-12 weeks)	Lactating does and suckling young	Pregnant does but not lactating	Adults at maintenance (bucks etc)	Complete feed for reproduction & fattening
Digestible energy	Kcal/kg	2500	2600	2500	2200	2500
Metabolizable energy	Kcal/kg	2400	2500	2400	2120	2410
Fat	%	3	3	3	3	3
Crude fibre	%	14	12	14	15 - 16	14
Indigestible crude fibre	%	12	10	12	13	12
Crude protein	%	16	18	16	13	17
Amino acids						
Lysine	%	0.65	0.75	-	-	0.70
Sulphur amino acids	%	0.60	0.60	-	-	0.60
Tryptophan	%	0.18	0.22	-	-	0.20
Threonine	%	0.55	0.70	-	-	0.60
Leucine	%	1.05	1.25	-	-	1.20
Isoleucine	%	0.60	0.70	-	-	0.65
Valine	%	0.70	0.85	-	-	0.80
Histidine	%	0.35	0.43	-	-	0.40
Arginine	%	0.90	0.80	-	-	0.90
Phenylalanine + tyrosine	%	1.20	1.40	-	-	1.25

Contd

Minerals						
Calcium	%	0.40	1.10	0.80	0.40	1.10
Phosphorous	%	0.30	0.80	0.50	0.30	0.80
Sodium	%	0.30	0.30	0.30	-	0.30
Potassium	%	0.60	0.90	0.90	-	0.90
Chloride	%	0.30	0.30	0.30	-	0.30
Magnesium	%	0.03	0.04	0.04	-	0.04
Sulphur	%	0.04	-	-	-	0.04
Trace elements						
Iron	mg/kg	50	100	50	50	100
Copper	mg/kg	5	5	-	-	5
Zinc	mg/kg	50	70	70	-	70
Magnesium	mg/kg	8.5	2.5	2.5	2.5	8.5
Cobalt	mg/kg	0.1	0.1	-	-	0.1
Iodine	mg/kg	0.2	0.2	0.2	0.2	0.2
Fluorine	mg/kg	0.5	-	-	-	0.5
Vitamins						
Vit. A	iu/kg	6000	12 000	12 000	6000	10 000
Vit. D	iu/kg	900	900	900	900	900
Vit. E	mg/kg	50	50	50	50	50
Vit. K	mg/kg	0	2	2	0	2
Vit. B1	mg/kg	2	-	0	0	2

Vit. B2	mg/kg	6	-	0	0	4
Pantotheneic acid	mg/kg	20	-	0	0	20
Vit. B6	mg/kg	2	-	0	0	2
Vit. B12	mg/kg	0.01	0	0	0	0.01
Niacin	mg/kg	50	-		-	50
Folic acid	mg/kg	5	-	0	0	5
Biotin	mg/kg	0.2	-	-	-	0.2

ANNEXURE-VI

Haematogical and biochemical parameters of rabbit

Haematological parameters of rabbit	
RBC (x10^6/mm^3)	6 – 10
Hb (g/dl)	10 – 16
RBC (10^3/mm^3)	3 – 11
Neutrophils (%)	10 – 42
Lymphocytes (%)	50 – 95
Eosinophils (%)	0 – 4.5
Monocytes (%)	0 – 3
Basophils (%)	0 – 1
Platelets (x10^3/mm^3)	200 – 500
Biochemical parameters of rabbit	
Protein (g/dl)	4.5 – 6.2
Calcium (mg/dl)	60 – 150
Urea nitrogen	12 – 25
Bilirubin (mg/dl)	0.25 – 0.6
Sterol (mg/dl)	25 – 135

ANNEXURE-VII

Organ Weight of Rabbit in Relation to Body Weight

Organ	Organ weight (g)	Organ weight (as % of body weight)
Body	3352	100
Head	178	5.3
Trunk	1855	55.3
Forelimbs	122	3.64
Hindlimbs	481	14.3
Integument	517	15.4
Musculature	1501	44.8
Digestive tract	128	3.8
Liver	103	3.1
Pancreas	6.47	0.19
Lungs	16.9	0.50
Heart	9.0	0.27

Contd.

Spleen	1.52	0.04
Thyroid	0.29	0.008
Adrenals	0.59	0.017
Kidneys	22.7	0.67
Brain	11.7	0.35
Spinal cord	5.8	0.17

ANNEXURE-VIII

Disease	Cause	Symptoms	Prevention	Treatment
Abscesses	Bacterial infection	Enlargements under skin near jaw. Can occur on other parts of body where there are wounds or scratches.	Minimize fighting. Eliminate sharp objects that can injure the rabbit.	Clip the fur around abscess, then lance and remove the pus. Disinfect with peroxide and use an approved antibiotic ointment or powder to promote healing.
Caked Mammary Gland	Milk production is in excess of utilization. Too few young or the young are not nursing.	One or more of the mammary glands may become swollen, hot and firm. Often mistaken for mastitis.	Reduce ration by one-half the day the doe kindles and gradually increase to full feed in 7 days.	Relieve some congestion by partialmilking. Massage gland with an anti-inflammatory ointment.
Coccidiosis	Five different species of protozoa *(one celled animals)* can injure the bile ducts, intestines or cecum.	Affects primarily young rabbits. Insevere cases, symptoms are diarrhea, loss of flesh, pot belly, loss of appetite,rough fur coat.	Keep pens clean and prevent fecal contamination of feed and water. Use wire floor pens.	Sulfaquinoxaline in food or drinking water continuously for 2-weeks.
Conjunctivitis "Weepy Eye"	Inflammation of eye lid. Can result from irritation by dust, sprays, fumes or by bacterial infection.	Excessive tear formation and fluid runs down cheek. Rabbit rubs eyes with itsfront feet which further aggravates the condition. Rabbits with snuffles frequently have "weepy eye".	Minimize sources (dust, dirt) of irritation. Remove chronically affected rabbits.	Use a commercial eye-washing product to remove dust, dirt or other foreign bject. If inflammation persists, an ophthalmic ointment containing antibiotics may be helpful.

Contd.

Ear Canker	Ear mite	Scabs or a crust start forming at base of inner ear. Mites cause considerable irritation and rabbit will shake its head and try to scratch ears with hind feet.	Do not let unaffected rabbits in contact with rabbits that have ear canker.	Use a cotton swab to apply mineral, vegetable or olive oil over all visible crust.Usually one or two treatments is sufficient.
Enteritis:Diarrhea	Likely caused by coccidiosis or other intestinal inflammation. In young rabbitsexcessive "greens" will cause diarrhea	A watery diarrhea may be only symptom.	Develop a program to control coccidiosis. Do not feed "greens" torabbits.	Keep rabbits comfortable and encourage feed consumption. A medicated or vitamin fortified feed may be helpful.
Mucoid	Thought to result from an irritant, a toxin or other stress factors such as dietarychanges, antibiotics and travel stresses.Not thought to be contagious.	Characterized by sub-normal body temperature, loss of appetite, toothgrinding depression, rough fur coat,dehydration, bloated abdomen anddiarrhea containing mucus material.	Prevent or reduce stress factors and have strict sanitation of hutches, feeding and watering equipment.	No successful treatment known, butmortality may be slightly reduced byfeeding a medicated or vitamin fortified food. Eliminate all affected rabbits, andpurchase breading stock that are free of the disease.
Eye Infection	Several types of bacteria.	The eyes of baby rabbits may stick shut and pus around the eyes may be noticed	Prevent cold drafts and other stress conditions.	Apply an antibacterial ophthalmic eyeointment.
Fur Block "Hair Ball"	Rabbit ingests a large amount of its ownhair or that from another rabbit in a shortperiod of time.	Rabbit will sporadically eat small amounts of feed. A firm mass can be palpated in the stomach.	Fur block is noted more often in Angora rabbits.	An oral dose of 1/2 oz. mineral oil may be effective. If the fur block is too large,surgical removal is necessary.

Contd.

Fungus Infection	Fungus	A dry scaly skin (dandruff) on the shoulders and back.	Prevent contact with affected rabbits.	Apply tolnaftate to affected area. Othermedications are available.
Heat Exhaustion	Excessively high temperatures (above 850F) and high humidity (above 70%),obesity, poor ventilation, insufficient water and crowding.	Increased respiration rate (panting) prostration, and excessive saliva discharge. Pregnant does and young in thenest box are most susceptible.	Provide shade, adequate air movement,and plenty of cool water on a hot day.Wet burlap in cage will aid in cooling.	Rabbits suffering from heat exhaustioncan be immersed in lukewarm water toreduce body temperature to the normal101-1040F. Apply a cool compress tothe ears.
Hutch Burn	Usually associated with wet and dirtyhutches. Urine and fecal material causethe skin to become irritated and infected.	Usually, a brownish crust covers theinfected area and a bleeding exudate may be seen.	Do not allow fecal material to accumulatein the hutch.	Clean affected skin with a germicidalsoap and keep the area clean.
Ketosis	Obesity, large litter, lack of exercise.	Usually occurs just before or just after kindling. Does go off feed and will not eat.	Prevent young does from getting too fat and make sure the does are eating at kindling time.	Obesity can be prevented by limiting feed to 4 to 6 oz. daily for does.
Mastitis "Blue Breast"	A bacterial infection in the mammarygland caused by an injury from nest box, cage or a bite from nursing young.	The mammary gland will become red incolor, swollen, tender and dark blue streaks may appear.	Reduce any chance of injury tomammary gland as the doe enters the nest box. Do not transfer the young to another lactating doe.	Penicillin is effective against staphylococcus or streptococcus organisms.
Malocclusion "Buck Teeth"	An inherited condition characterized by excessive	Lower teeth protrude and upper teethcurve into the	Do not use rabbits with "buck teeth" for	Teeth on young rabbits can be trimmed until they reach

Contd.

	growth of the front teeth.	mouth.	breeding purposes.	slaughter weight.
Pastaurellosis "Snuffles"	A chronic upper respiratory ailmentcaused by bacterial infection.	Typical early signs are a nasal discharge,watery eyes, head shaking, sneezing or rattling noise in breathing anda loss of weight. Continual infection maycause rabbits to become sterile.	Quarantine all new rabbits for a 3 week period. Cull rabbits that show continuous chronic symptoms of thedisease. Reduce stress factors and have good management and sanitationpractices.	Tetracycline in feed may be effective inan uncomplicated case. For accuratetreatment, the organism needs to becultured and identified.
Pneumonia	A bacteria or virus infection usually associated with other respiratory diseases or stress factors such as damp,drafty and unsanitary hutches.	Elevated temperature, labored breathing,nasal discharge and loss of appetite aresymptoms.	Control of "Snuffles" reduces chances of pneumonia. Eliminate drafts and wet conditions.	Treatment may vary depending upon the causative agent. A broad spectrum antibiotic may be helpful.
Ringworm	A fungus infection that can appear onany part of the body but most often willoccur on the head. This fungus infection can be transmitted to humans.	Loss of hair in circular patches. A yellowish crust forms and sloughs off inabout 3 weeks. If infection goes untreated, considerable scratching willoccur.	Do not allow infected rabbit in contact with other rabbits. Wear gloves whenhandling the infected rabbit.	Clip the fur around affected area and treatwith toinaftate. Other medications areavailable.
Salmonellosis "Scours"	Several members of the Salmonella groupof bacteria can affect rabbits.	General symptoms are diarrhea, loss ofweight, conjunctivitis, and rapid breathing.	Good husbandry practices and aquarantine of new rabbits.	Tetracycline in the feed may be effective

Contd.

Skin Mange	An infestation of mites that burrow through the skin and cause considerable irritation. Mange mites that affect dogs and cats can infect rabbits.	There will be some loss of hair, scaly skin and intense itching and scratching.	Infected rabbits should be separated and treated. The premises should be disinfected.	Dust with an insecticide approved for the purpose. Treatment should be repeated in 7-days.
Sore Hocks	A bruised or infected area on the undersurfaceof the hock joint. Related stress factors are a dirty hutch, rough surfaceson the floor, or a floor constructed with the wrong size wire-grid.	Scabs usually appear on the bottom of the rear feet. Pad and toes of the front feet may become infected. Nervous andheavy rabbits are more susceptible to the sore hock condition.	Do not let wet litter and manure accumulate. Eliminate sharp objects on the floor.	Wash the hock with a germicidal soap andapply an antiseptic such as tincture ofiodine.
Spirochetosis "Vent Disease"	Infection by a spirochete and can betransmitted by mating.	Blisters or scabs occur on the sex glands.	Always check the breeders before mating. Isolate all new rabbits.	Injection of penicillin is the best treatment. Do not breed until all lesionsdisappear.
Tyzzers Disease	A bacterial infection usually associated with stress conditions and poorsanitation.	The acute form results with diarrhea, dehydration and death within 48 hours.	The best preventative measures are clean stock and good husbandry practices	Treatment with antibiotics is not very effective.
Warbles	The growth of the botfly larvae under theskin.	Warbles can be observed as enlargementsunder the skin with a small hole in thecenter. The condition is painful and the rabbit may lick the area.	Screen the area to keep the botfly awayfrom the rabbit.	Remove the larvae without crushing it. Clean the wound with a germicidal soapor apply an antibiotic powder orointment to prevent infection.

❑❑❑

Index

a

Abortion 95
Acute renal failure 105
Aggressive behaviour 155, 157
Amputation 133
Angora fur 16
Angora rabbit 137, 138
Anorexia 127
Arthritis 134
Autoimmune 90

b

Bacterial 86, 107
Breeds 11, 17

c

Caecal impaction 115
Caecotrophy 45
Cage identification 78, 139
Cage system 25
Cages with litter 27
Cages without litter 27
Calcivirus 124
Californian cage system 28, 30
Candidiasis 83
Cannibalism 95
Carbohydrates 50
Castration 75
Catherization 147
Cheek teeth 41
Chronic diarrhea 122
Chronic renal failure 105
Classification 11, 12, 14, 16
Coccidiosis 125, 126
Colibacillosis 116, 120
Colony cages 27
Combing 140
Commercial 12
Compact 12, 13
Compact battery 28, 32
Complementary diet 54
Components of rabbit diet 48
Coronavirus 123
Creep diet 52
Curing 162
Cylindrical 12, 13
Cystitis 101

d

Dental formula 41
Depilation 140
Determining pregnancy 72
Dietary enteritis 123
Digestive glands 44
Digestive system 39, 111
Digestive tract 39, 45

Diseases of liver 128
Diseases of pancreas 130
Diseases of rabbit 81
Diseases of reproductive system 94
Diseases of spleen 129
Domestic rabbit breeds 17
Dressing 159, 162
Drinkers 33, 34
Dystocia 95

e

Ectoparasites 83
Endometrial hyperplasia 96
Endometritis 97
Endotracheal tubation 151
Enteritis 115
Enteritis complex 122
Enterocolitis syndrome 120
Enterotoxemia 119
Equipments for cages 33
Examination 144
Extensive reproduction rate 67

f

Fancy rabbit 16
Fear aggression 158
Feed regimes 52
Feed troughs and racks 35
Feeding behaviour 47, 155, 157
Feeding of rabbit 55
Feeding schedule 53, 54
Female reproductive system 60
Fertilization 63
Fibre 48
Flat deck system 28
Food aggression 157
Fostering 73
Fractures 131
Full-arch 121
Fungal & yeast infestations 82

g

Gastric intestinal obstruction 113
Gastric stasis 111
Gestation 53, 63, 138
Glossing 162, 163
Glycosuria 105
Gonad development 60, 62
Grading 141, 163, 164

h

Haematuria 101
Hairballs 112
Handling 143
Handling rabbits 76
Harvesting of wool 139
Hind leg paralysis 136
Hind leg paresis 135
Housing rabbits 21
Housing systes 24
Humidity 22
Hydrometra 97

i

Ideal environmental conditions 21
Identification 77, 138
Incisors 41
Inclined slope battery system 28, 31
Individual cages 26
Individual identification 78
Indoor cages 26
Induction 150
Inhalation anaesthesia 149
Injection 147

Insemination 65
Intensive reproduction rate 67, 68
Intussusception 113

k

Kindling 64, 73, 138

l

Lactation 53, 65
Large intestines 44
Lead poisoning 128
Light 21
Lighting 66

m

Male reproductive system 59
Mastitis 98
Maternal behaviour 155, 156
Mating 71, 137
Minerals 50
Miscarriage 95
Miscellenous 89
Mites 85, 84
Mouth 39, 111
Multi colour fur 16
Myositis 135

n

Neoplasia 93, 99, 135
Neoplasm 93
Nest box 35, 37, 38
Normal fur 14
Nutrient requirements 55

o

Obesity 128
Oesophagus 43
Oestrous cycle 62
Oral medication 146
Osteomyelitis 132
Osteoporosis 134
Outdoor cages 26
Ovulation 63

p

Packing 162
Pain aggression 158
Pasturellosis 107
Perineal urine scalding 104
Pet rabbit 17, 97, 128
Pharynx 43
Pin worm 124
Plucking 140
Post-traumatic osteomyelitis 132
Pre-anesthetic preparation 147
Pregnancy toxaemia 94
Premedication 148
Preservation 160, 161
Problems of parturition 95
Production cycle 71
Protein 49
Pseudopregnancy 64, 94
Puberty 60, 62
Pyometra 97

r

Rabbit behaviour 155
Rates of reproduction 67

Record keeping 79
Red urine 100
Rex fur 15
Ringworm 82
Rotavirus 123

s

Salmonellosis 121, 129
Satin fur 15
Season 66
Sedation 148
Semen collection 64
Semi-arch 12
Semi-intensive reproduction rate 67, 68
Sexing 138
Sexual aggression 158
Sexual behaviour 155, 156
Shearing 139
Skin diseases 82
Slaughtering 159
Sludgy urine 101
Small intestines 43, 111
Social behaviour 155, 156
Sorting 163
Sperm production 61
Spinal cord injury 133
Splay leg 135
Sticky bottom syndrome 49
Stomach 43, 111
Stomach worm 124
Storage 160, 162

t

Tapeworm 124
Teeth 7, 40
Temperature 21, 66, 81
Territorial 155, 157
Territorial aggression 157
Textile 16
Ticks 86
Tyzzer's disease 121

u

Urinary incontinence 104
Urolithiasis 103
Uterine prolapse 96

v

Ventilation 23
Viral 89, 110, 123
Vitamins 50

w

Weaning 74, 138
Weaning diet 52
Weaning enteritis 117
Wild rabbit breeds 17, 18
Wool fur 16
Wool grading 141
Wool quality 140
Wool type 16

❑❑❑

FSC
www.fsc.org
MIX
Papier | Fördert
gute Waldnutzung
FSC® C083411

Zeitfracht Medien GmbH
Ferdinand-Jühlke-Straße 7
99095 Erfurt, Deutschland
produktsicherheit@kolibri360.de